NDT Methods Applied to Fatigue Reliability Assessment of Structures

SPONSORED BY
Fatigue and Fracture Reliability Committee
Structural Engineering Institute (SEI) of the American Society of Civil Engineers

EDITED BY
Jamshid Mohammadi

ASCE

SEI
Structural Engineering Institute

Library of Congress Cataloging-in-Publication Data

Non-destructive test (NDT) methods applied to fatigue reliability assessment of structures / prepared by Fatigue and Fracture Reliability Committee, Structural Engineering Institute, American Society of Civil Engineers ; edited by Jamshid Mohammadi.

p. cm.

Includes bibliographic references and index.

ISBN 0-7844-0742-8

1. Non-destructive testing. 2. Materials--Fatigue. I. Mohammadi, Jamshid. II. Structural Engineering Institute. Fatigue and Fracture Reliability Committee. III. American Society of Civil Engineers.

TA417.2.N654 2004

620.1'12--dc22 2004051812

Published by American Society of Civil Engineers
1801 Alexander Bell Drive
Reston, Virginia 20191
www.pubs.asce.org

ISBN 0-7844-0742-8
Manufactured in the United States of America.

Preface

This publication presents a series of papers on the application of non-destructive test (NDT) methods in fatigue and fracture reliability analysis of structures. The publication is intended as a collection of case studies where NDT methods have demonstrated to be an effective tool in the assessment of structural reliability. These case studies cover a variety of problems and provide the reader with in-depth discussions on very specific applications of the NDT methods. The Fatigue and Fracture Reliability Committee (FFRC) of the Structural Engineering Institute (SEI) of the American Society of Civil Engineers initiated this publication as a project through supports from SEI. FFRC has an active membership of 13-15 members at any given time. The project would not have been possible without the many hours of volunteer work by these FFRC members. The FFRC membership is pleased with the quality of papers presented in this publication and hopes that the papers presented herein can provide practicing engineers with an introduction of NDT methods with immediate applications to real world problems.

Fatigue and Fracture Reliability Committee

Contents

An Overview of Non-Destructive Test Methods in Fatigue and Fracture Reliability Assessment

Jamshid Mohammadi[1], MASCE

Abstract

With the growing demand for repair, retrofit and reconstruction of aging structures, there is a critical need for methods that can effectively and reliably be used in identifying potential modes of damage and the current condition of a given structure. When the applied load is of cyclic nature, an affected structure will be subject to fatigue damage. The decision to repair, retrofit or reconstruction of such a structure requires a careful planning and investigation on the nature and extent of damage and estimation of the remaining useful life of the structure. The estimation of damage and prediction of the remaining life can only be obtained through a probabilistic analysis using information on the applied stress history, fatigue and fracture behavior of the structure and the geometry of its critical components. Non-destructive test (NDT) methods play an important role in gathering such information and in conduct of a reliability analysis for determining the condition of the structure and estimating its remaining useful life. Depending on the type of a structure and its condition, NDT methods may simply involve an inspection process with data collection, especially on stress ranges at fatigue-critical components. More comprehensive methods may require a rigorous program of monitoring along with various tests and analyses to accurately assess the condition of the structure. The results of such techniques and analyses can be useful in establishing the fatigue reliability profile of the structure and in a cost-effective scheduling of future repair, retrofit or reconstruction activities. In this paper, an overview of NDT methods with specific applications to fatigue and fracture reliability estimation is presented. Specifically, this paper serves as an introduction to what is presented elsewhere in this volume on the subject of NDT methods to fatigue and fracture reliability assessment of structures.

---

1 Dept. of Civil and Arch. Engrg., Illinois Institute of Technology, Chicago, Illinois 60616

Introduction

Fatigue and fracture constitute major modes of failure to a large class of structures. The failure process often occurs in systems that are subject to a dynamic cyclic load action. Although the failure mechanism in essence is the same for various types of affected structures, the load process and the methods via which structures are assessed for failure may be different. The assessment usually begins with the evaluation of the current condition. This is then followed by a more rigorous assessment process that may range from visual inspection to X-ray inspection or detection of acoustic emissions (Byers, et al, 1997). These methods may also involve detailed structural analyses as well as long-term monitoring using sensors to compile data on applied stresses at fatigue- or fracture-critical components.

As it is well recognized, any such assessment programs will ensure the prolonged life of the structure and will be effective in preventing sudden failures. Various assessment methods, whether involve monitoring, analysis, and structural tests, are considered non-destructive if they do not interfere with the integrity and performance of the structure. A specific non-destructive test (NDT) method can be designed to address the structure's demand for fatigue and fracture reliability assessment. For lower-risk and less important structures, simple methods are often used; whereas, high-risk structures demand more detailed procedures that may require advanced NDT methods (Byers, et al, 1997). As expected, many NDT methods are available today; and newer methods are arriving every day. With such a variety of methods available, the decision on selecting a method for a given structure may become a challenging task. Successful past applications and case studies would offer an insight into whether a specific method is suitable for condition assessment in a given structure especially when fatigue and fracture modes of failure are prevalent.

The computation of fatigue damage for use in a reliability analysis is either based on the Miner damage rule or the fracture mechanics method utilizing crack growth behavior and stress cycle history. For either method, one needs to have information on the stress history, geometry, material strength, the dead load, and the overall fatigue damage or crack growth behavior of the component. The computation of the reliability is conducted for a given number of stress cycle and initial crack size (when the fracture mechanics method is used). For this computation, one of several common reliability analysis methods, such as the first or second order reliability method (FORM or SORM), or Monte Carlo simulation process may be used (see for example, Ang and Tang, 1984; Haldar and Mahadevan, 1999). The results are often presented in the form of a "reliability profile" curve, which shows the relation between reliability and stress cycles (or time). Regardless of the method of analysis selected for the system reliability estimation, the knowledge of uncertainty inherent in the system (or component) geometry, strength, fatigue or fracture behavior and stress history is needed. Various NDT methods may especially be helpful in providing data for use in the system reliability analysis.

To compile a series of sample application studies, the Fatigue and Fracture Reliability Committee (FFRC) of Structural Engineering Institute (SEI) initiated a project to compile a collection of papers that address the specific application of various NDT methods in fatigue and fracture reliability assessment. The papers presented elsewhere in this publication each focuses on a specific application and provides the reader with detailed information on how such methods can effectively be applied in real world problems. Applications cited discuss such problems as steel girder bridges, airframe systems, welded connections, bridge components (including concrete slabs) and other systems that may be subject to fracture failure. Although these applications are limited, they are intended to provide the reader with information on an overall process that may be used in similar situations.

An Overview of Various NDT Applications

As it is evident from several papers presented in this volume, structural monitoring dominates the NDT methods in fatigue and fracture reliability assessment. Structural monitoring is cost-effective and provides for the information needed for structural life estimation of structures. In some applications (such as in highway bridges), monitoring is applied for a relatively short period of time (a few days); whereas, in others (such as in aircraft) monitoring is often carried out for a much longer time (as long as several hundred cumulative flight hours). A discussion on the duration of data acquisition process for monitoring several types of structures (such as highway and railroad bridges, off-shore platforms, aircraft and structural members affected by vibrations generated by mechanical systems) can be found in Mohammadi (2001). The reliability assessment can be obtained only after an ample amount of data is compiled and a rigorous analysis is conducted to arrive at life estimates. The process of data compilation requires installation of sensors at various locations (where fatigue damage is more critical). Often a careful planning and evaluation of the structure will be needed to optimize the number of sensors and the data collection effort. Several of the application papers presented in this volume discuss the NDT application using the monitoring process (see papers by Fu; Mohammadi, et al; and Braun and Mohammadi).

In some applications, the process of condition assessment requires identification of the extent of damage, size of crack and deterioration of structural members. Non-destructive methods in these applications may involve using various available techniques such as acoustic emission, electro-magnetic, portable hardness test, liquid penetrant, magnetic particle, radiography, residual stress by hole drill, and ultrasonic methods. Each method is suitable for one or more problem areas such as local damages that cause discontinuity in the cross section and the material in the affected structure. Some of these methods may require use of test equipment and application of a force on the structure via an impact-producing device. The information from these methods can be used in identifying how the resistance capability of a component has been reduced. This in turn can be used along with a detailed structural analysis to determine the changes that have occurred to the resistance against fracture, fatigue and cracking. This volume contains papers that address these methods in applications that may involve fracture of materials such as concrete used in bridges (see the paper by Son and Mohammadi).

Welded connections are often considered to be among the fracture-critical components of a structure. Fracture in welds in structural members is affected by stress cycles as well as the existence of initial discontinuities in the weld materials and the geometry of the welds and the connections. Non-destructive test methods for reliability assessment of welded connections may include X-ray techniques along with advanced structural analysis models such as the finite element method. There is also an uncertainty involved in the detection of discontinuities in the weld. This uncertainty must also be included in the reliability assessment. Initial cracks may also be a source of fatigue and fracture failure in connections using fasteners. In this regard, the uncertainty in detecting the initial crack and the potential for a multi-site damage condition need to be considered in the reliability assessment and development of an NDT method that will be suitable for the structure (see paper by Mahadevan and Zhang). Two papers, in this volume, present methods related to NDT of welds and uncertainties in detecting damage (see papers by Sarkani, et al; and Sarkani and Michaelov). Analytical methods using finite element analysis and NDT methods utilizing X-ray techniques are explained and provide the reader with information on how such techniques can be applied in fatigue and fracture reliability assessment.

In conducting an NDT for condition assessment, often the availability of new information needs to be incorporated in the assessment process. New information may be the result of a new data acquisition process; use of newer techniques in assessing the extent of damage in members; and in some cases can be the result of actual failure cases. A method to incorporate the new information in the assessment process will be through the statistical updating process. The procedure in detecting the crack size can be improved by conducting NDT along with an updating process. An application along this line and the corresponding model formulations and results are presented in this volume (see paper by Zhao and Haldar).

Reliability Assessment Process

In nearly all applications presented in this volume, the reliability assessment process involves analytical methods following the NDT process. In fatigue-critical structures, the assessment process is conducted by using the NDT data to determine the probability that a component will surpass a designated fatigue life. Computation of fatigue damage and fatigue reliability is based on Miner's damage rule and uncertainties inherent in the damage formulation utilizing FORM, compiled data and component's fatigue characteristics (Ang and Munse, 1975; Byers, et al, 1997). The results are presented in the form of a reliability profile that indicates the reliability versus the number of load cycles. The profile can also show the outcome of any inspection and/or repair done on the system. Usually, a repair will result in an increase in reliability; even though the increased reliability will never reach the system reliability when it first started its service. This method of fatigue reliability assessment is explained in this volume in several application papers that deal with bridge structures. In fracture-critical components, the method of analysis is usually based on theories of fracture mechanics. The reliability is estimated based on the probability that the component will surpass a designated number of stress cycles for a specific crack size. Again, FORM is often used for this purpose. The process depends on the structural detail and the existence of an initial crack. Those applications that involve fracture reliability analysis in airframe systems and consideration for the multi-site damage utilize this type of analysis. Reliability profiles can also be established using this method showing the reduction in reliability in terms of the number of stress cycles along with changes in reliability resulting from any repair and/or improvement made to the components. Once the reliability profile is established, it can be used for the decision making process regarding the structure's demand for future inspection, repair or replacement. With the availability of additional data, the reliability profile can be updated using a Bayesian approach (Ang and Tang, 1975) as demonstrated by Zho and Haldar in this volume.

Expected Benefits of NDT-based Fatigue Reliability Assessment

The cost associated with the NDT process and the reliability assessment varies depending on the method employed. In any application, this cost can be incorporated into an overall life-cycle cost analysis of the structure. Through a comprehensive life-cycle cost analysis, one can determine whether the use of a particular method is justified. Considering all the expenses associated with the operation and maintenance of a fatigue- or fracture-critical structure, one may decide to conduct a simple NDT approach at earlier life of the structure and develop strategies to implement more advanced methods at later stages. A program of routine inspection and maintenance can also be considered as a safeguard against sudden drops in the system fatigue and fracture reliability. This may involve structural monitoring on a short term or continuous basis. Major benefits of a comprehensive NDT approach incorporated into the life-cycle cost analysis of the structure include a prolonged service life, a more controlled overall cost for operation and maintenance; and an effective method in preventing sudden failures. A reliability profile can be used to optimize the intervals between inspections and the overall maintenance, condition assessment and repair cost.

Summary and Conclusions

This volume presents a collection of papers on the application of NDT methods in fatigue and fracture reliability assessment. In general, and as an overview on the NDT methods applied to fatigue and fracture reliability assessment, the following conclusions are noted:

- Non-destructive methods used in fatigue and fracture reliability often involve structural monitoring;

- Depending on the application, NDT may range from simple visual inspections to more advanced methods utilizing various devices to estimate the extent of damage to the affected structure.

- Selection of a particular NDT method depends on the type of structure and its importance in terms of risk for failure.

- An effective plan for the NDT process can be incorporated into the structure's overall life-cycle cost analysis.

- Methods used for fatigue and fracture reliability analysis use the NDT results along with a follow-up detailed structural analysis.

References

Ang, A. H.-S. and Munse, W. H. (1975). "practical reliability basis for structural fatigue." *Preprint No. 2494, Proc., National Structural Engineering Conference*, ASCE, New York, NY.

Ang, A. H.-S. and Tang, W.H. (1975). *Probability concepts in engineering planning and design, Vol I-Basic Principles*, John Wiley and Sons, New York, NY.

Ang, A. H.-S. and Tang, W.H. (1984). *Probability concepts in engineering planning and design, Vol II*, John Wiley and Sons, New York, NY.

Byers, W.G., Marley, M.J., Mohammadi, J., Nielsen, R.J., and Sarkani, S. (1997). "Fatigue reliability reassessment procedures: state-of-the-art paper." *J. Struct. Engrg.*, ASCE, 123(3), 271-276.

Byers, W.G., Marley, M.J., Mohammadi, J., Nielsen, R.J., and Sarkani, S. (1997). "Fatigue reliability reassessment applications: state-of-the-art paper." *J. Struct. Engrg.*, ASCE, 123(3), 277-285.

Haldar, A. and Mahadevan, S. (1999). *Probability, reliability, and statistical methods in engineering design*, John Wiley and Sons, Inc., New York, NY.

Mohammadi, J. (2001). "Field data requirements for fatigue reliability analysis of aging structures," *Proc., Structures Congress 2001*, P. Chang, editor, American Society of Civil Engineers, Reston, VA.

Structural Monitoring as a Non-Destructive Test Method in Fatigue Reliability Assessment of Aging Aircraft

Joseph F. Braun[1] and Jamshid Mohammadi[2], MASCE

Abstract

This paper presents the application of a continuous program of structural monitoring in fatigue reliability assessment of aging aircraft. The monitoring system consists of sensors installed at fatigue critical locations of an airframe and compiling data on the flight load history and stress range fluctuations. The data are then used along with a detailed damage analysis to arrive at fatigue reliability of the structure and an estimation of the remaining fatigue life. The paper discusses the procedure in compiling flight load data; the duration of the data acquisition period and description of flight parameters that can effectively be used in fatigue reliability analysis. The validation process for the compiled data and the method of analysis for the estimation of the remaining fatigue life are explained.

Introduction

Flight data monitoring has proven to be an effective method for the evaluation of structural integrity and estimation of fatigue life expended (FLE) of aircraft. A successful monitoring program requires a comprehensive effort involving data gathering, analysis and engineering evaluation and interpretation of all parameters that influence the aircraft load conditions. The flight parameter data in the form of time history and/or frequency of occurrence of peak and valleys are required for the development of load spectra for fatigue analysis of critical structural components of aircraft.

----

1 Systems & Electronics, Inc., 190 Gordon Street, Elk Grove, Illinois 60007
2 Dept. of Civil and Arch. Engrg., Ill. Inst. of Tech., Chicago, Illinois 60616

During a flight operation, an immense amount of data on flight parameters is gathered so that nearly all possible values of the key parameters, that influence the load conditions of the aircraft, can be captured. This calls for an advanced system capable of: (1) handling the volume of incoming flight data; and (2) processing data on a real-time basis. Recent developments in advanced monitoring systems provide the structural engineer with the ability to collect a massive amount of data pertaining to an aircraft's loading, strains and flight parameters. Most current systems are capable of providing in-flight and post-flight data analyses to various degrees and qualities. While the in-flight real-time analysis provides a maximum level of data compression, it compromises data visibility. This means that the user will not be able to examine the in-coming data until it has been processed and displayed by the data acquisition system. Thus unless a quality assurance program is built-in the system, the accuracy of the data cannot be readily verified. Furthermore, any adjustment in the data compilation criteria cannot easily be made. This system is, of course, providing a rapid analysis of the data. Its major drawback is that if there is a mistake in data processing criteria, the data may be totally lost without a chance of full recovery. The post-flight data analysis procedure, on the other hand, provides for a maximum data visibility but limits timely data error detection. However, in cases where mistakes in data processing parameters are made, almost the entire data are recoverable by adjusting the processing method. Generally, the post-flight data analysis scheme requires more effort and time in producing the final processed flight data. An ideal data acquisition system is a combination of the in-flight and post-flight processes. Such a system provides the user with an adequate data visibility and verification of data accuracy. Furthermore, a convenient format can be sued to ensure speedy feedback brought about by rapid data access. This idea was utilized by the authors in compiling flight data on a variety of applications. The system used is ASH-37 developed by Systems & Electronics, Inc. (SEI). The success of ASH-37 is in its ability to offer a versatile monitoring system that can readily be configured to suit the aircraft specific data acquisition requirements.

The type of data needed for fatigue reliability evaluation of an airframe system depends on the airframe design and the availability of structural fatigue parameters that describe damage growth. To a large extent, most fixed-wing aircraft require flight data in the form of peak and valley exceedance statistics. Such data provide information on the frequency of peak and valley of stress and normal load data in batches containing several hundred hours of flight operation. When damage tolerant (crack growth) analysis is required, the exceedance data can readily be used along with the information on the stress intensity factor and geometrical parameters of critical components. The outcome of this analysis will be the probability that a potential crack will grow to a predetermined size. This probability is then used as the basis for estimating the structural reliability, and the time between inspection and/or repair. In cases where crack initiation analysis is of concern, the exceedance data can still be used. In such a case, the information on the fatigue life versus stress range values in the form of Goodman diagrams is needed. The analysis follows the Miner-Palmgren damage rule to estimate the probability that a critical component will fail for a pre-determined number of stress cycles. This information can then be used in estimating structural reliability and the time between scheduled maintenance and/or repair.

In rotary-wing aircraft, damage analysis requires data compilation in the form of maneuver and sequence of events during flight operations. ASH-37 can easily be configured to comply with data acquisition demand in this case. The flight maneuver and sequence of event recording provides a detailed information on aircraft various operations during flight. This also allows for recording of various key flight parameters at each flight condition at each event. Development of a built-in data quality assurance procedures will then provides for a means to verify the expected ranges of flight parameters. To a limited extent, this procedure is effective in identifying problem areas that may encounter during the data acquisition process. The processed data can then be used as an input to the development of a comprehensive structural integrity evaluation. The item-by-item sequence of events data recording is especially helpful in damage assessment in a more refined manner by identifying flight conditions that are more severe to the aircraft condition. Most flight conditions are predetermined in this type of data acquisition method. However, if the recorder detects an unidentified condition persistently occurring, it will automatically establish a new condition and sequence of events.

While the focus of fatigue life investigation of fixed-wing aircraft is mainly on structural components, in rotary-wing aircraft mechanical and dynamic components are extensively investigated for their useful life and reliability. And as such, the data for rotary-wing aircraft, to a large extent, is compiled for these components. Often, the data comprises of the amount of time an aircraft spends in a particular loading regime (as described earlier). This data is essential in estimating the time at which a component must be replaced. The estimation of the life to replace a component follows a similar procedure as that in the damage accumulation method, which is used in crack initiation analysis. Thus the type of field data in rotary-wing aircraft is different from that in fixed-wing aircraft which primarily use stress range data in a damage tolerant analysis of structural components.

This paper presents an overview of the ASH-37 monitoring system and its application as an aid in a non-destructive evaluation of fatigue reliability of airframe structures. The paper describes the data compilation process; data compression and analysis; data quality assurance and fatigue reliability analysis. The applications cited in this paper were primarily used in conjunction with structural integrity analysis of the U.S. Navy aircraft. The paper also discusses the selection of aircraft flight parameters that are critical to aircraft loading conditions; and each parameter specific requirements in terms of data collection and processing. The discussion focuses on both fixed-wing and rotary-wing aircraft.

Operation Usage and Flight Parameters Monitoring

The selection of flight parameters for use in a comprehensive structural monitoring program depends on the type of aircraft and the configuration of its structural components. Furthermore, the selection of desired parameters depends on the flight condition for which the data are being compiled. Unlike civil engineering structures that often need a few parameters for monitoring, airframe structures are generally very demanding in terms of the number of parameters required in a comprehensive monitoring program. Accordingly, systems capable of handling data acquisition and storage of data for a large number of parameters will be needed in airframe structures. Furthermore, such systems must offer a light weight and the ruggedness needed in aircraft flight loading conditions. These loading conditions often constitute rapid fluctuations in the normal load factor. Normal load factor is often defined by the acceleration component N_Z acting in the vertical direction as the aircraft maneuvers. Flight parameters selected for fixed-wing (e.g. A4 and P-3) and rotary-wing (e.g. CH-46 and AH-1W) aircraft for monitoring are described below. These parameters were used by authors in their investigations.

Fixed-wing Aircraft -- A complete flight history of aircraft will be required such that data for various load levels experienced by fatigue-critical components may adequately be established. In a typical application, the fixed-wing aircraft (e.g. A4) parameters selected for monitoring consist of: (1) strains at several fatigue-critical locations; (2) normal acceleration (N_Z); (3) longitudinal acceleration (N_X); (4) roll acceleration; (5) indicated airspeed; (6) pressure altitude; (7) fuel quantity; (8) weight-on-wheels (WOW); (7) stores release; and (8) in-air refuel. In conducting a fatigue reliability analysis, one needs to know the history and distribution of the strains and load factors. All the other parameters are needed for several reasons that are indirectly involved with the fatigue analysis. The information on the fuel quantity is needed in computing the changes in gross weight. The gross weight affect the normal load which in turn is correlated with stress levels at structural components. The information on airspeed and altitude and most other parameters are crucial in identifying various flight conditions that may particularly trigger higher strains on components. In addition, these parameters identify whether the compiled data follow the norm in a particular aircraft.

The data acquisition, data compression and a limited amount of analysis are conducted during the flight. The algorithm for data sampling rate is introduced to the recorder to maximize the storage capacity. The occurrence of peaks and valleys in the load data determines the severity of fatigue damage experienced by the aircraft components. The load data are obtained as the product of the gross weight W and N_Z. Since fatigue damage accumulation is affected by the number of load cycles, it is logical to use the peak and valley occurrence as a basic criterion in recording all other flight parameters including strains. The recorder is pre-configured to conduct the sampling using the peak and valley cycle counting and process data for temporary storage in the memory. The system is configured to record only the data that are significant to fatigue damage. In all other times, where data are not significant, the recorder will switch off to a *sleep* mode of operation. This conserves energy and optimizes the data storage capacity. The criterion for peak and valley occurrences can be based on any desired cycle-counting approach. General procedures for cycle counting can be found in Broek (1989) or in ASTM (1987). The method known as the "rainflow" is used to count cycles based on the occurrence of a rise followed by a fall or a fall followed by a rise. If the variation of the parameter (N_Z) is within a value that causes stresses below the fatigue endurance limit of the component, no cycle is counted. This limit (specified by a ± value) is referred to as the *deadband*. The cycle-counting results are arranged in tabular or graphics forms showing the frequencies of peaks and valleys at various N_Z levels (e.g., 0.5*g*, 1.0*g*, 1.5*g*, ..., or -0.5*g*, -1.0*g*, -1.5*g*, etc., in which *g* is the gravity acceleration). Figure 1 shows a typical peak/valley counting procedure used by ASH37 recorder. Table 1 summarizes the recording of peaks and valleys depicted in Fig. 1. The peak and valley data can also be generated for strain levels. This can be done for all possible values of the gross weights or for different gross weights and different flight conditions. A flight condition is often designated with a *flight purpose code* (FPC). An FPC determines the loading condition and the ranges of key flight parameters. Table 2 presents a sample of output from ASH37. Figure 2 presents an actual peak and valley data processed for a critical structural component in the wing of a fixed-wing aircraft. Notice that in Fig. 2, there were 40 valley counts at -1000 micro in./in. strain. At 1000 micro in./in. strain, the recorder had about 180 peak counts. For the aluminum member, this strain value is equivalent of 1,000 psi (7 MPa) stress. The record shows 6 peak occurrences of strain at 2000 micro in./in. This corresponds to 2,000 psi (14 MPa) stress.

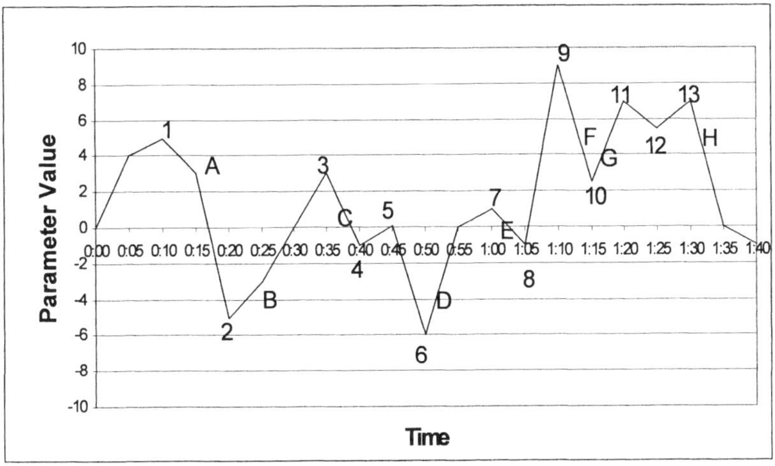

Fig. 1 Peak and Valley Designation

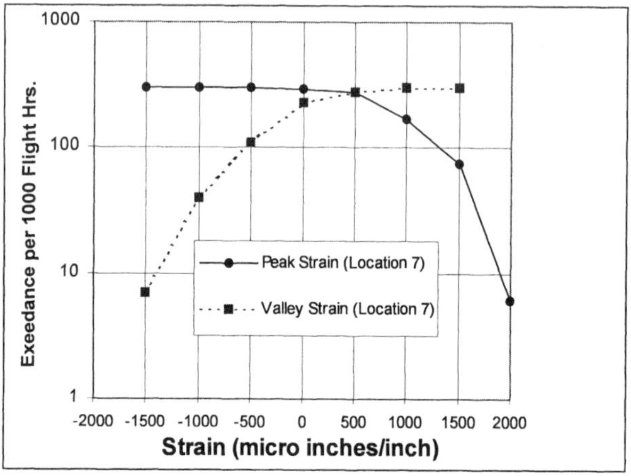

Fig. 2 An Example of Peak and Valley Exceedance Data

Table 1 Algorithm for Recording Peaks and Valleys Depicted in Fig. 1

Point & Occurrence	Logic
1 -- Peak above	Meets rise/fall criterion at A, above deadband.
2 -- Valley below	Meets rise/fall criterion at B, below deadband.
3 -- Peak inside	Enters deadband from below, meets rise/fall criterion at C.
4 -- Invalid valley	Meets rise/fall criterion but previous peak was inside deadband, successive peaks and valleys inside deadband are *not* recorded.
5 -- Invalid Peak	Meets rise/fall criterion but previous valley was invalid (inside deadband).
6 -- Valley below	Meets rise/fall criterion at D, below deadband.
7 -- Peak inside	Enters through lower boundary, is valid at E.
8 -- Invalid valley	Meets rise/fall criterion but previous peak was invalid (within deadband).
9 -- Peak above	Meets rise/fall criterion at F, above deadband.
10 -- Valley above	Meets rise/fall criterion at G, above deadband.
11 -- Invalid peak	Does not meet rise/fall criterion.
12 -- Invalid valley	Does not meet rise/fall criterion (since Point 11 is not valid).
13 -- Peak above	Value is larger than that at Point 11, meets rise/fall criterion at H.

Table 2 Sample Output from ASH37 Recorder

FILE	FLIGHT	REC. NO.	TIME	RECORDS
N20B1700.dbf	01	00023	1.36	1 to 275

Record #	INDCTR	PARAM	TIME	A/S	Nz	Strain 1
17	V	5	413.51	166	0.89	238
18	P	5	423.26	166	1.19	381
19	V	5	430.32	166	0.89	191
20	P	5	467.29	166	1.19	429
21	V	5	473.36	166	0.89	286
22	P	5	479.99	166	1.29	476
23	V	5	489.68	166	1.09	381
24	P	5	486.44	166	0.89	238
25	C	0	512.57	166	1.19	381
26	V	5	529.88	166	0.99	286

V=vally, P=peak, C=continuous record, A/S=airspeed

The data shown in Fig. 2 can be directly used in fatigue reliability analysis as explained later.

Rotary-Wing Aircraft -- In the CH-46 application, twenty parameters were selected for the survey. These parameters were: normal acceleration (N_Z); roll angle; pitch angle; heading angle; rotor RPM; outside air temperature; longitudinal speed; rudder pedal position; weight-on-wheels; rotor brake; collective stick position; airspeed; altitude; fuel quantities (left and right); cruise guide indicator; lateral speed (doppler); engine torque (left and right); cargo hook strain; longitudinal stick position; and lateral stick position. The fatigue damage characteristics of various critical components of the aircraft depend on values of many of these parameters. Damage indices are generally developed by the manufacturer. These indices indicate the level of damage experienced by a given component in terms of the occurrence of flight parameters within certain pre-determined values. The normal acceleration is measured via an advanced system called *motional pick-up transducer* (MPT). This sensor is a stand-alone device used along with the ASH37 recorder. The outside air temperature was obtained from a probe installed on the forward left side of the aircraft. The external load was obtained from a strain sensor mounted to the external cargo hook. Most other parameters were acquired using the aircraft gages. A similar set of parameters was also selected for data acquisition in the AH-1W application.

For certain structural components aboard rotary-wing aircraft, usual methods of fatigue damage assessment may still be used. However, as indicated earlier, the manufacturers often use a series of damage indices that require compilation of the flight load and other flight parameters in a special format. In the applications cited in this paper, the data in the form of sequence of events and load regimes were compiled. In this format, several specific sequence of events (described in terms of ranges for flight parameters) are defined. The recorder will compile data once it detects that one or more of these events have occurred.

Processing Flight Data

The processing of flight data is partially performed by the recorder. Additional analyses are then performed upon downloading the data onto a desk-top computer.

In fixed-wing applications, the triggering factor in recording data is the occurrence of a peak or valley in either the N_Z data or the strain data. The recorder has a built-in criterion to recognize the occurrence of a peak or valley for data recording. In addition, several additional analyses and data compression tasks are conducted during the flight. For example, in most applications, the frequency of various levels of the normal acceleration (N_Z) encountered during the flight is desired. Such data include the number of times a specific N_Z level has been identified to either experience a peak or valley. The recorder can easily be configured to change the cycle counting method as may be required by a specific application. The recorder generates the compressed data in tabular form, ready for further post-flight processing. An indicator warns the crew of full capacity. At such time, the recorder can easily be downloaded to a ground processing unit for further analysis. Usually, upon data downloading, a rigorous evaluation of the data is conducted to ascertain the accuracy and robustness of the flight data compiled. This process is explained later in this section. Post-flight data analysis includes development of cycle-counting stress or acceleration data; and development of input for fatigue reliability evaluation. In addition to recording of peak and valley cycles during a flight, the recorder also compiles data on the number of take-offs, landing and the occurrences of *weight-on-wheels* and *touch-and go* cases. These occurrences are also important in assessing fatigue damage to some components of the aircraft.

In rotary-wing applications, all input parameters to the recorder are monitored for identification of various sequence of events and specific maneuvers experienced by the aircraft during the flight. The sequence of events that define a maneuver is determined from the flight spectrum specified by the aircraft manufacturer. For example, in the AH-1W application, a total of 293 uniquely definable flight conditions were incorporated in the recorder data processing algorithm. Each defined flight condition was, in turn, separated into three gross weight ranges. These were: the low weight range defined by a gross weight of less than 5,760 kg (12,500 lbs); the medium range defined by a gross weight of between 5,760 and 6,125 kg (12,500 and 13,500 lbs); and, the heavy weight range defined by those in excess of 6,125 kg (13,500 lbs). Table 3 presents two samples of the sequences of events that uniquely identify a specific maneuver in the AH-1W application. A similar process is also adapted in the CH-46 application.

Table 3 Samples of Sequence of Events in Flight Data Acquisition of AH-1W Aircraft

Maneuver	Parameter Range	Sequence of Events
Normal Turn @ 0.7 V_h	$1.3g \leq N_Z < 1.8g$ $0.55V_h \leq$ A/S $< 0.75V_h$ Roll altitude >|10°|	Roll exceeds 10° angle of bank. N_Z Exceeds 1.3g but is below 1.8g. All parameter ranges are recorded.
Straight And Level @ 0.5 V_h	$0.5g \leq N_Z < 1.3g$ $0.4V_h \leq$ A/S $< 0.55V_h$ Roll altitude <|10°|	N_Z goes back to below 1.3g. Roll goes to <10° angle of bank. All parameter ranges are recorded.

A/S = airspeed, V_h = threshold airspeed

The ASH-37 recorder executes the maneuver recognition algorithm on a real-time basis starting from aircraft power up to power down. The final product at the end of each operation is a complete history of the flight including: the time of occurrence of each maneuver; the time spent in each maneuver; and the values of all parameters that show a change. Furthermore, the weight ranges in which the maneuver occurs is identified and marked. If an unidentified flight condition is encountered, the recorder automatically assigns an *unrecognized mode* recording. Simultaneously, the recorder logs a record of each flight parameter band; thereby compiling the data on the ranges for the parameter. Each time a parameter value changes into a new band, the system logs the parameter that experienced the change and the corresponding new band value. This type of recording continues until one of several pre-identified built-in *recognized* flight conditions is encountered. The data compiled for an unrecognized mode of recording provide a detailed description of those flight parameters that are significantly affected. If an unrecognized mode is repeated over a period of time, the system will treat the mode as a new flight regime and will add it to the aircraft's flight spectrum.

Data Quality Assurance Procedures -- To ascertain the accuracy and robustness of the incoming and compressed data, a rigorous quality assurance (QA) program is implemented. This program consists of a host of activities that range from simple checking of a randomly selected set of data to a complete set of statistical tests to determine trends in the data and establishing confidence levels. Again, some of these activities are conducted during the flight; while many others are post-flight activities. The in-flight quality assurance activities are intended to detect cases of malfunction in sensors or the recorder. For example, if the recorder measures the same stress level for a prolonged time period, one may suspect that a gage may have experienced a *slip*. The recorder also has error detection signals for proper operation and maintenance. Most post-flight QA activities are focused on conducting statistical tests on the data to verify that data variations are within accepted ranges. The variability in flight data may be because of:

- Uncertainties associated with the method of data acquisition process. Marginal errors in predicting various thresholds and bands for flight parameters are among uncertainties inherent in the method of data acquisition process. In most applications, the estimates for a flight parameter bands are derived based on either the aircraft's previous flight records or records of identical aircraft. In any case, such information is subject to variability.

- Errors associated with sensor sensitivities.

- Marginal errors encountered during installation of sensors, initial settings, tests and calibrations.

The QA procedures are intended to provide a safeguard against recording erroneous data that do not fall within the expected norms. Although a total elimination of such errors is not possible, certain measures can be taken to minimize their occurrence and recurrence during the data acquisition period. The QA procedure used by authors in the aircraft flight data acquisition is conducted in three steps. The first step is a set of in-flight activities conducted by the system as *self-tests*. The second step consists of post-flight statistical analyses of the data to establish data trends and to identify any discrepancy, in the data, that may be an indication of certain deviations from the norms. The final step involves the identification of the source of any discrepancy in the data and the development of an appropriate corrective measure to remedy the potential problem areas. Figure 3 presents a schematics diagram of the QA activities. A brief description of the three steps involved in the QA procedure is provided below.

Initially, the recorder is introduced with several pre-identified flight conditions. In fixed-wing applications, usually, the deadbands for key parameters and typical ranges for all parameters are introduced into the system. The ranges offer the first line of defense against collection of erroneous data. In rotary-wing applications, flight conditions are the maneuvers and sequence of events obtained from the manufacturers. If data compiled consistently show trends outside of the designated conditions, the prevalence of a problem areas is suspected. Discrepancies are then flagged for further investigation. If the recorder does not recognize a flight condition, with all parameters being within their respective acceptable ranges, a new flight condition will be added. This is only after a careful investigation reveals that there are no anomalies associated with the data acquisition process. This constitutes Step 1 in the QA process.

As flight data are compiled, batches of data are randomly selected for detailed statistical analyses. The purpose of these analysis is to investigate: (1) any statistical correlation that may exist between two or more parameters; (2) a trend or pattern in the distribution of a parameter or group of parameters; (3) consistency among blocks of data compiled for a given flight parameter for the same aircraft or for several identical aircraft; and (4) consistency of the data when compared with published results, design values and norms indicated in the aircraft's manual. An inventory of qualified data or published results are maintained for reference. The reference data is periodically updated and revised, if necessary, as new data sets become available. Dramatic changes occurring in the data are carefully investigated to identify and remedy potential problem area associated with the data acquisition process. This constitutes Step 2 in the QA process.

Corrective measures are taken only if a problem area has been identified and proven to be the source of error or dramatic change in the data. Appropriate measures could consist only of a few adjustments in the previously-selected thresholds for a parameter; a re-calibration of sensors; an added feature to the previously-identified flight conditions; a re-configuration of the system; or replacement of the recorder. In most applications the replacement of a sensor or the entire data acquisition system is rarely needed. Most often the problem area can be corrected with a few changes in the system configuration. This constitutes Step 3 in the QA procedure.

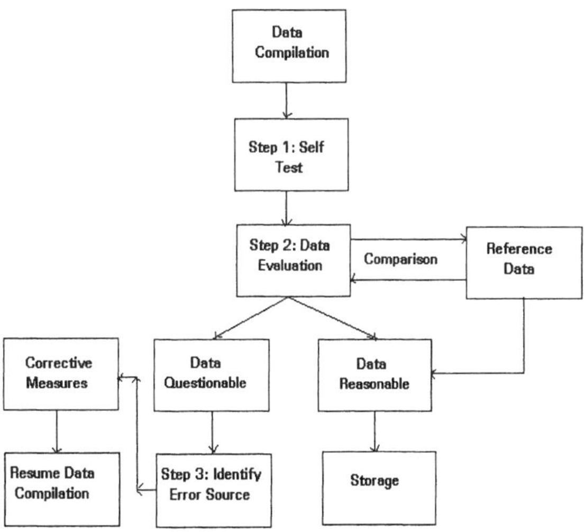

Fig. 3 Quality Assurance Procedure for Compiled Flight Data

Fatigue Reliability Assessment

The results of aircraft flight surveys are used in estimating the fatigue life expended (FLE) and assessment of fatigue reliability of critical components of the structure. The data acquisition process through the on-board recorder (such as ASH-37) offers a very efficient cost-effective non-destructive technique for these applications.

In most rotary-wing applications, the data are provided to the aircraft custodian for further fatigue reliability analysis. The procedure via which the reliability assessment is conducted depends on the type of aircraft. Usually, a series of damage indices are developed that are used along with the sequence of events and load regime data to arrive at estimates for FLE. When FLE estimates exceed certain critical values for a component, the component is then scheduled for inspection or replacement.

In Fixed-wing applications, the reliability assessment process is more generic and follows two general methods. These are: (1) the crack initiation method; and (2) the crack propagation (damage tolerant) method. The details for both methods are well-established and extensively reported in the literature (see for example Barsom and Rolfe, 1987). In crack initiation method, the statistics of the strain peak and valley are converted to the stress range values. Damage occurred to the component is estimated using the Miner-Palmgren Damage rule. If there are k different ranges of stresses are identified in the compiled data, the damage D is estimated as

$$D = \frac{n_1}{N(S_1)} + \frac{n_2}{N(S_2)} + \dots + \frac{n_k}{N(S_k)} \tag{1}$$

In which n_i is the number of cycles for the stress range S_i (i=1 to k); and N (S_i) is the number of cycles to failure if stress range is equal to S_i and is computed from the S-N relationship for the component for the stress range S_i. The S-N relationship is written in the following form

$$N=C/S^m \tag{2}$$

In which C and m are constants. The S-N relationship is developed for different components for different dead load stresses. Failure in a component occurs when D=1. The family of S-N curves generated is often shown in the form of the modified Goodman diagram. Based on the S-N relationships for a component, curves representing plots of constant fatigue lives (in terms of the number of stress cycles to failure) are shown in a two-axis system in which the horizontal axis shows minimum; and the vertical axis the maximum stress. Each curve represents the locus of all combinations of maximum and minimum stress at which failure will occur in the indicated number of cycles. Thus for any combination of minimum and maximum stress, the corresponding fatigue life can be obtained. Figure 4 presents the general format of this type of diagram. The parameter R in the figure, is the ratio of the maximum to minimum stress. The ultimate stress values are taken as 120 ksi, as an example, in the figure. Notice that for R=-1, the largest stress range is obtained; and as such, the fatigue life is very short. Diagrams similar to Fig. 4 are essential for estimating FLE. By computing the damage D from Eq. 1, one can determine the FLE index based on the number of peak and valley stresses compiled through flight data acquisition. Through an extrapolation, the approximate flight hours until failure (i.e., D=1) can be estimated and used as a means for scheduling a maintenance or part replacement. In practice, usually a probabilistic method is used for fatigue reliability analysis. The probabilistic analysis is useful in incorporating the uncertainties inherent in: (1) the flight data; (2) the Miner-Palmgren damage rule; and (3) the S-N relationship. The outcome of a probabilistic analysis is often the estimation of the remaining useful fatigue life of a component based on an accepted probability of failure.

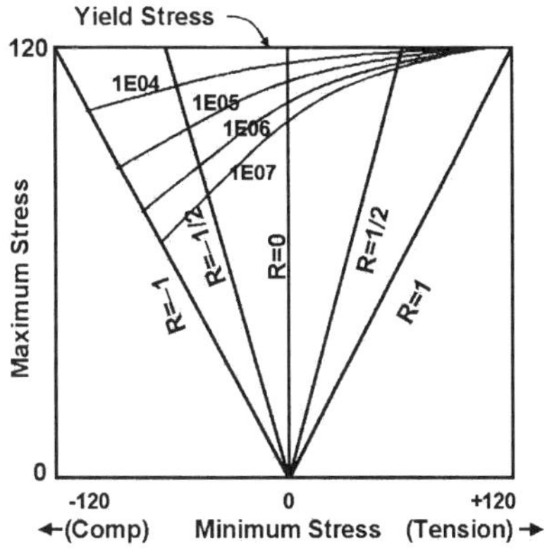

Fig. 4 Fatigue Life Data Presentation in Terms of Maximum and Minimum Stresses

In some applications the crack growth method is used for fatigue reliability analysis. The output from the ASH-37 recorder, in the form of peak and valley statistics, can directly be used in this type of analysis also. The crack growth analysis utilizes the following general equation (Barsom and Rolfe, 1987):

$$\frac{da}{DN} = A(\Delta K)^m \tag{3}$$

Where a=crack length; N=number of stress cycles; ΔK=stress intensity factor fluctuation; and m is a constant. As the number of stress cycles advances, the crack will grow according to Eq. 3. The parameter ΔK depends on many factors including the geometry of the structural detail, and material properties. In general, ΔK is written as a function of the crack length a and stress level S. In a simple form (see Fig. 5) where a crack is through the thickness of the metal

$$\Delta K = c_0 \Delta S \sqrt{2\pi a} \tag{4}$$

In which c_0 is a constant. In practice, an appropriate form of Eq. 4 is obtained and used for the critical component in estimating the length of crack for various stress cycles N. Most often, one needs to perform a numerical computation to solve Eqs. 3 and 4. For most aircraft fatigue-critical components, the information on the stress intensity factor can be found in Gallagher, et al (1984). Equations 3 and 4 can also be treated in a probabilistic formulation incorporating the uncertainties inherent in the stress range ΔS, stress cycles and in the crack growth model represented by Eqs. 3 and 4. Using the probabilistic analysis, one can estimate the probability that the crack will grow to a predetermined size after a target length for the flight hours is reached. The computation of the structural reliability utilized by the authors was based on a first order reliability method (FORM). The FORM was used in establishing a matrix of values for the reliability versus flight hours for various potential crack sizes. Given a desired reliability level (say 99%), the matrix was used to establish graphs showing crack growth versus flight hours.

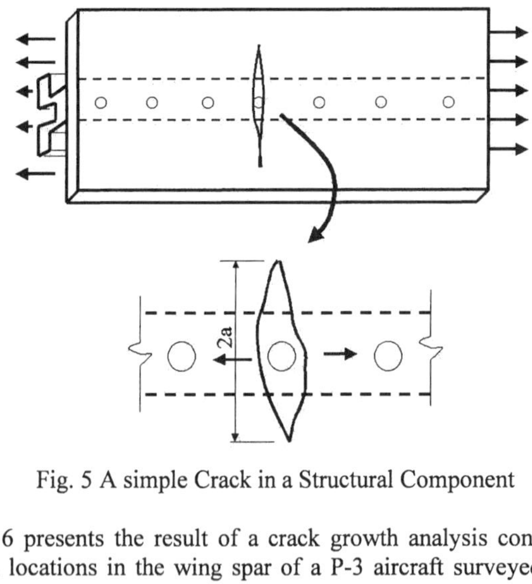

Fig. 5 A simple Crack in a Structural Component

Figure 6 presents the result of a crack growth analysis conducted for three fatigue critical locations in the wing spar of a P-3 aircraft surveyed by Systems & Electronics, Inc. (SEI). The flight parameters included several strains at three wing locations identified by wing stations WS188, WS167 and WS 205. The crack growth analysis utilized a probabilistic method extrapolating crack growth (Eqs. 3 and 4) for future potential stress values developed in the three stations if the aircraft loading condition continues the same as those compiled in the SEI flight data acquisition of the P-3 aircraft. The number of stress cycles used in the analysis was converted into actual flight hours for presentation of the results in Figure 6. As seen in this figure, at around 15,000 flight hours, the crack size will grow to nearly 0.2 inches (10 mm). The fatigue lives reported in Fig. 6 correspond to a 99% reliability level. Considering this as a target critical size for the components, we can expect a thorough inspection or possibly even replacement of the parts.

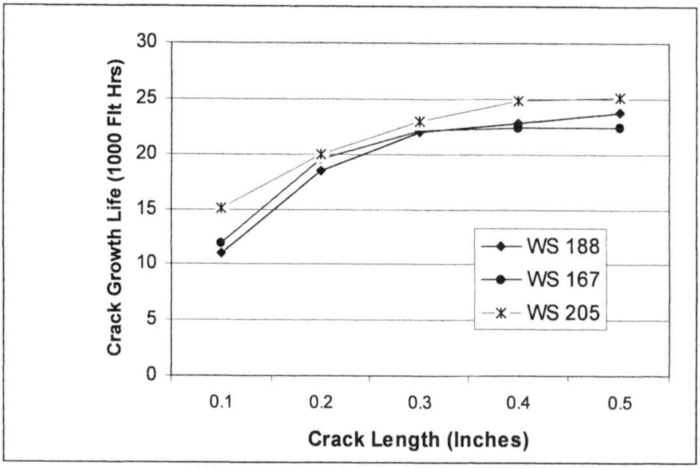

Fig. 6 Example of Crack Growth Results

Summary and Conclusions

The process of flight data acquisition for aircraft is explained in this paper as a non-destructive means for fatigue and fracture reliability assessment. The paper describes various steps in this process and the type of flight data compiled in conjunction with fatigue reliability analysis. For fixed-wing aircraft, the data acquisition process is primarily focused on peak/valley occurrences of key flight parameters such as the normal acceleration and strains. For rotary-wing aircraft, the data presented in three weight groups and summarize the time spent in each of many pre-identified flight conditions (loading regimes). The fatigue reliability assessment can be conducted from flight data based on the fatigue damage characteristics of individual critical components. Major conclusions of this paper are:

- A continuous monitoring of flight parameters provides an effective non-destructive method for tracking the structural integrity of aircraft.

- The compiled data can easily be used along any of the two methods of fatigue analysis.

- The item-by-item sequence of events data recording in rotary-wing aircraft is helpful in providing a more refined and descriptive way of summarizing flight data.

- The self-test and automatic upgrading of flight data conducted by the air-borne data acquisition system are features that are especially helpful in assuring the quality of the incoming flight data.

- The data acquisition system offers an efficient method for compiling many hours of flight data in one load. An ample amount of information on the aircraft parameters is acquired and summarized by the system in each flight.

References

ASTM (1987). *Annual book of ASTM standards*, Section 3, Volume 03.01, Designation: E 1049, American Society for Testing of Materials, Philadelphia, PA.

Barsom, J.M. and Rolfe, S.T. (1987). *Fracture and Fatigue Control in Structures, applications of fracture mechanics,* Second Edition, Prentice Hall, Englewood Cliffs, NJ.

Borek, D. (1989). *The practical use of fracture mechanics,* Kluwer Academic Publishers, Amsterdam, the Netherlands.

Gallagher, J.P., et al (1984). *USAF damage tolerant design handbook*, Rept. #AFWA L-TR-82-3073, University of Dayton Research Institute, Dayton, OH.

Nondestructive Testing for Steel Highway Bridges

Gongkang Fu [1] Member ASCE

Abstract

Non-destructive testing can be used to examine and/or monitor steel highway bridge members for welding quality control and/or detecting possible fatigue failure. This paper presents a brief overview on the subject. It covers several non-destructive testing methods, spanning from those routinely used to some still under development. These methods may be classified into two groups: 1) for global diagnosis and 2) for local diagnosis. Global diagnosis methods refer to those that concentrate on the entire bridge structure system. Their typical result is a local area of suspected discontinuity, damage, or failure. In contrast, local diagnosis methods are those applied to a local area. Their typical result is identification of the damage or failure with the details quantified, such as the crack's size, shape, direction, etc. As seen, local diagnosis methods provide much more conclusive diagnosis. However, for field inspection of steel bridge components, local diagnosis methods have to be applied to local areas identified using global diagnosis methods.

Introduction

Steel highway bridges may be subjected to a large number of high stress cycles over their life of service, especially those on heavily traveled roadways. These bridges can be susceptible to fatigue failure. Of them, those designed and constructed according to earlier design and construction specifications are further more vulnerable because these codes did not have adequate provisions regarding the fatigue failure risk.

----

[1] Professor and Director, Center for Advanced Bridge Engineering, Department of Civil and Environmental Engineering, Wayne State University, Detroit, MI 48202, gfu@eng.wayne.edu

Small, invisible discontinuities in the steel material have been considered to be the origin of visible fatigue failure (cracking) in steel bridge components. Accordingly, it is important to prevent fatigue failure in the process of fabricating these components by detecting those small discontinuities introduced. In addition, it is often even more critical to identify early fatigue failure in bridges in service. In these activities, non-destructive testing techniques play an important role, simply because no destruction (or at lease no significant destruction) is permitted when inspecting a bridge component, be it in the fabrication process or in service condition.

Thus, non-destructive testing for identifying discontinuity is the objective in assuring the safety of steel highway bridges against fatigue failure. In the process of steel bridge component fabrication, application of non-destructive testing is relatively simpler because the members are easily accessible. Certain techniques (e.g., X-ray testing) can even be applied in an automated manner (using scanning). Therefore detecting discontinuity focuses on a local, small area at a time. Accordingly, these testing methods are referred to as local diagnosis methods, although scanning a large number of such local areas can cover a large area.

The situation becomes very different when inspecting bridge components under service condition. It is because the members to be inspected may be difficult to access. Providing direct access to all these members may become extremely costly. Accordingly, inspection of bridge components in service often takes a two-step approach. The first step is to identify those local areas that are more likely to have fatigue failure. This step requires an examination of the entire bridge structure system. Thus, it is referred to as global diagnosis here. If no suspected areas are identified as a result of global diagnosis, the inspection is considered completed. Otherwise, local diagnosis will be needed to individually focus on each of the identified area for conformation and quantification of the damage. This second step is referred to as local diagnosis hereafter.

Global and local diagnosis methods are respectively discussed below in this paper. Some of these methods are fairly mature and are applied routinely. In contrast, some of them are still under development as pointed out. Since the local diagnosis methods are used for both the shop and field conditions, they are discussed first.

Local Diagnosis Methods

Many of service-induced problems in steel bridge-components are directly related to discontinuities due to inherent structure, processing, or fabrication of the steel members. Fatigue cracks may originate from these discontinuities and then propagate further, which may ultimately affect the safety of the bridge structure by causing more significant discontinuity. Nondestructive test methods discussed below are used to locally detect, verify, and quantify cracks in steel bridge components. The interested reader is referred to [FHWA-HI-88-006 1986, ANSI/AASHTO/AWS 1988, AASHTO 1991], for a more comprehensive coverage.

Liquid Penetrant Testing (PT)

PT is mainly used to identify or confirm steel fatigue cracking in the field. The basic procedure of PT requires the test area be pre-cleaned to remove all surface contaminants. When applied to a surface of steel material, the penetrant liquid will seek out and enter small cracks. Excessive penetrant is then removed from the test surface by wiping or water rinsing. After drying, a developer is applied. The penetrant remaining in the cracks bleeds out, forming a highly visible contrasting indication on the test surface. PT relies on ability of the penetrant liquid to enter into a discontinuity. Therefore it can only find cracks that are open to the surface of the material. It actually can be applied to other materials than steel provided that they are non-porous and not adversely affective to the penetrant liquid material. PT is simple and inexpensive, very portable and well adapted for field use, and can be very sensitive. On the other hand, if discontinuities (cracks) are filled with contaminants, rust, oxidation, or corrosion products, they may not be detected. PT also requires good visual acuity of the operator and its sensitivity is reduced at low temperatures. More details about PT can be found in [FHWA-HI-88-006 1986], including various available penetrant liquids for respective applications, detailed procedure requirements for surface preparation, application of penetrant and developer, interpretation and evaluation of indication, etc.

Magnetic Particle Testing (MT)

MT is mainly used to identify or confirm steel fatigue cracking also in the field. MT requires steps of surface preparation, magnetization, particle application, removal of excessive particles, result interpretation, and post cleaning and demagnetization. It causes a magnetic field of sufficient strength and a predetermined direction to leak if discontinuities are present. Typical discontinuities detectable with MT include cracks, lack of fusion, and other weld related surface discontinuities. For example, AASHTO requires fillet weld repairs be test by MT for fracture critical members [AASHTO 1978]. More details about the test method can be found in [FHWA-HI-88-006 1986], including various methods for respective applications, commercially available equipment, procedure requirements, etc.

Radiographic Testing (RT)

RT exposes film to X or gamma radiation that has penetrated a specimen. It then processes the exposed film, and interprets the resultant radiograph by identifying possible discontinuities pictured. RT relies on penetration and absorption characteristic of X or gamma radiation to the steel material tested. Thus both sides of the tested component have to be accessible. This test method is more often used in the shop in the process of steel bridge component fabrication for welding quality control. When radiation passes through a material, source of the radiation is absorbed. The amount of absorption is dependent on thickness and density of the material being tested. A detector is used to record the variation in intensity of the emerging radiation beam. An important factor affecting ability to detect and record discontinuities is the position of the discontinuity. For best detection, the greatest dimension of the discontinuity should be parallel to the direction of radiation. The selection of RT is governed by the following factors. 1) The type and thickness of the steel material. 2) The geometry of the steel material. 3) The required sensitivity for detection. AASHTO specifications [1978] contain provisions to guide RT examination for steel bridge components.

Ultrasonic Testing (UT)

Sound has to be produced or transmitted through media, such as air, liquids, and solids. While the medium is changed to one to another, the way of sound propagation will be altered. Using ultrasonic waves and their propagation patterns in solid materials like steel, UT can evaluate the internal condition of these materials. Similar to RT, UT provides information about a material's internal properties. Typical discontinuities detectable by use of UT include lamination, cracks, and many of the surface- and subsurface-weld-related discontinuities. For example, the pin-and-hanger details have been examined using UT for cracks hidden behind accessible surfaces. Note that proper calibration is required for reliable diagnosis. This is because the component's geometry significantly influences the patterns of ultrasonic wave propagation, and it varies from bridge to bridge. The AASHTO specifications [1978] contain provisions to guide UT examination for steel bridge components.

There are four different ultrasound wave modes used for discontinuity detection and quantification. Longitudinal or compressive wave has the highest velocity, and particle vibration is parallel to the direction of wave propagation. Compression and shear wave modes are commonly used for weld testing. Surface- or Rayleigh-waves have shown some success for fatigue crack detection of in-service bridge welds, although their application is limited by the required smooth surface. Plate- or Lamb-waves are only used for very thin materials and therefore are not appropriate for bridge components.

UT is readily adaptable to field testing and portable units are available. It is also routinely used in the shop condition for steel bridge component fabrication. Some forms of cracking are best suited for UT. It is most successful for detecting discontinuities that are orientated perpendicular to the direction of propagating sound wave. It is therefore an ideal complementary method when used with RT. Thus combination of UT and RT may result in an optimum coverage. On the other hand, UT should not be used on rough surfaces, on parts with complicated geometry, on highly attenuative materials, or where the discontinuity size is expected to be smaller than one or half of the wavelength used. Note that application of UT requires proper training, and poorly written procedures have been a limiting factor for its successful application [FHWA-HI-88-006 1986].

Acoustic Emission Testing (AE)

AE is described as a method to detect deformation or flaw growth by detecting associated sound energy release. The most notable difference between AE and other methods is that an applied stress is required to cause flaw growth and the acoustic emission. AE monitors the actual condition of the component at the time of test, and this process cannot be repeated exactly, because flaw growth is an irreversible process. Thus, it has been applied in the field condition for steel bridge components in service. When acoustic emissions attributed to a load causing crack extension eventually cease, further emission will not occur unless the load is increased. Primary limitation of AE for bridge structures is that it is difficult to recognize the sound energy released by a growing flaw from background noise cased by, e.g., joint friction and traffic. Some AE applications avoid this problem by isolating known noise-generators. Due to this limitation, this method is still under research and has not been as widely used as the first four methods discussed.

Global Diagnosis Methods

Global diagnosis methods refer to those that concentrate on the entire system of steel bridge in the service condition, as opposed to a local area (e.g., a weld). The objective of global diagnosis methods is to identify local areas that may have fatigue cracking. Further confirmation or denial of such cracking may be required using one ore more local diagnosis methods discussed above, unless the crack is very obvious. Unfortunately, visual inspection has been the only routinely applied method for steel bridges for fatigue failure. On the other hand, intensive research is underway to develop more reliable and cost-effective non-destructive testing methods in this direction. This section briefly discusses these research efforts, as well as the visual inspection method.

Visual Inspection

As of now, visual inspection still plays an important role in general condition assessment for highway bridge structures and also in load tests for load rating discussed elsewhere [Fu 1999]. Efforts have been, and are still being, made to develop other global detection techniques for specific type(s) of damage (e.g. [Fu 1997a 1997b, Alampalli and Fu 1997]). It is not exaggerating to state that, at this time, visual inspection is the only reliable global diagnosis method available. Periodical bridge inspection mandated in the United States is largely performed with visual examination, for fatigue cracking ad other distress. Anomalities are detected by recognizing changes from normal conditions in color, shape, dimension, location, smoothness, and reflection to light, etc. Of course, professional experience is important in such inspection, particularly when exercising special attention in areas more vulnerable to certain failure risks. Some examples of fatigue crack identification can be found in [Yen et al 1990].

Research Efforts of Global Diagnosis for Highway Bridges

Developing non-destructive testing methods to improve visual inspection for global diagnosis has been a focus of research for the past two decades. Proposed methods have been overwhelmingly based on dynamic modal testing using accelerometers that measure acceleration when the structure is excited by an external force such as a truck [Doebling et al 1996]. These sensors can perform measurement at a limited number of points on the structure. Covering a large area using many accelerometers could be very expensive, and no techniques are available for effective diagnosis using measurements from only a few points [Fu 1997a, 1997b]. The effectiveness of using modal testing to diagnose (including locating) small fatigue damage needs to be established and quantified. This is because measurement data inevitably contain noise. This noise makes it difficult to identify the signal, especially when local and small damages are concerned. Some previously proposed approaches suggested to also include numerical modeling as a dominant or complementary tool (e.g., finite element modeling). Such modeling can also be costly for a large number of bridges in a transportation network. In addition, it remains unknown whether numerical modeling can realistically cover the inconsistency between the nominally specified and the as-built conditions. Examples of such inconsistencies include real concrete strengths being different from the specified values, non-structural members adding structural stiffness, etc.

Recently, a new approach to global diagnosis was proposed using high resolution images, which can be provided by a coupled current device (CCD or CCD camera) [Fu et al 2001a, 2001b]. A typical CCD is shown in Fig.1, which has a sensor to receive light signals to process to digital images. The sensor has an array of 2048 by 2048 sub-sensors referred to as pixels. This type of devices has the following advantages in acquiring needed data for damage diagnosis for structural safety monitoring. 1) No sensors need to be attached to the structure, which significantly reduces instrumentation costs including that for interrupting and interrupted normal operation of the structure. 2) A large number of points on the structure can be covered for which measurement data are to be obtained. This offers an unprecedented amount of spatially intensive data for enhancing diagnosis resolution and effectiveness. 3) The prices for CCD have become much affordable and are expected to be more so in near future, due to rapid advancements in the computer and electronics industry.

The pixels can be viewed as sensors for data acquisition purposes. These sensors provide information on the bridge structure's physical location, which can be used to obtain displacement, slope, and curvature of the structure. This set of information can be used as the bridge's structural signatures for diagnosis purposes.

Fig.1 A CCD Camera

Namely, when changes in these signatures are detected, it is understood as a signal of stiffness change or structural damage such as fatigue cracking.

The proposed approach in [fu et al 2001a, 2001b] uses a probabilistic concept for diagnosis, taking into account possible noise in data acquired experimentally. For a model steel structure in the laboratory, simulated fatigue damage was introduced for several levels of severity, ranging from 3 to 13 percent of stiffness reduction. Results of applying the proposed method show that all these damage levels were detectable. Field application of this method is still underway to understand its capability or sensitivity for realistic conditions of steel highway bridges in service.

Discussions and Conclusions

There exist reasonably reliable local diagnosis methods for nondestructively inspecting steel bridge components for discontinuity, including fatigue damage, in the shop and field conditions. Nevertheless, efforts are being spent to advance some of these methods and to develop more advanced methods. On the other hand, visual inspection is currently the only routinely practiced method for globally diagnosing steel bridge components for fatigue damage. In recent years, significant research efforts have been spent to develop more quantitative and cost-effective methods for non-destructive testing. These proposed methods have not reached a stage of routine application.

References

AASHTO Commentary on the ANSI/AASHTO/AWS D1.5-88 Bridge Welding Code, Washington, DC, April 1991

AASHTO Guide Specifications for Fracture Critical Nonredundant Steel Bridge Members, Washington, DC, 1978

AASHTO Manual for Condition Evaluation of Bridges. Washington, DC 1994

AASHTO Standard Specifications for Welding of Structural Steel Highway Bridges. Washington, DC, 1981 (3rd ed.)

S.Alampalli and G.Fu "Signal versus Noise in Damage Detection by Experimental Modal Analysis", *ASCE J. Structural Engineering,* Feb.1997, Vol.123, No.2, p.237

ANSI/AASHTO/AWS D1.5-88 Bridge Welding Code, An American National Standard, 1988

S.B.Chase "Developing NDT Technologies for the Next Century", *Structural Materials Technology*, Proc. Of An NDT Conference, (Ed.) by P.E.Hartbower and P.J.Stolarski, San Diego, CA Feb. 20-23, 1996

Doebling, S.W., Farrar, C.R., Prime, M.B., and Shevitz, D.W. "Damage Identification and Health Monitoring of Structural and Mechanical Systems from Changes in Their Vibration characteristics: A Literature Review", *LA-13070-MS Report,* Los Alamos National Laboratory, May 1996.

FHWA-HI-88-006 Nondestructive Testing Methods for Steel Bridges, Participants Training Manual, FHWA, USDOT, June 1986

G.Fu, A.G.Moosa, and J.Peng "Optical Nondestructive Evaluation for Structures" (in press) *Research in Nondestructive Evaluation*, 2001

G.Fu, A.Moosa, and J.Ye "Structural Damage Diagnosis Using High Resolution Images" (in press) *Journal of Structural Safety*, 2001a

G.Fu and A.G.Moosa **"**An Optical Approach to Structural Displacement Measurement and Beyond**"**, (in press) *ASCE Journal of Engineering Mechanics*, 2001b

G.Fu "Nondestructive Bridge Testing", Chapter 6, *Bridge Safety and Reliability: Concepts and Code Developments*, ASCE Committee on Safety of Bridges, ASCE, 1999

G.Fu "Bridge Inspection: Modal Testing for Global Diagnosis", *Proc. 7th International Conference on Structural Safety and Reliability,* Kyoto, Japan, (Ed.) by N.Shiraishi, M.Shinozuka, and Y.K.Wen, Nov. 24-28, 1997a, p.475

G.Fu “Issues in Nondestructive Testing for Global Diagnosis of Bridges” *Proc. Joint ASME/ASCE/SES Summer Meeting*, Evanston, IL, June 29-July 2, 1997b, p.27

Inst. of Structural Engineers "Report on the Testing of Structures", UK, Sept. 1964

Inst. of Structural Engineers "Appraisal of Existing Structures", UK, July 1980

T.Javor "Testing Bridges in Situ", Materials and Structures, RILEM, Vol.9, No.53, 1976, pp.369-374

K.H.Klippstein and P.A.Romito “Variable Amplitude Load Fatigue: Task A - Literature Review”, *FHWA-RD-87-060*, July 1990

Ontario Ministry of Transportation and Communications "Ontario Highway Bridge Design Code", Downsview, Ontario, Canada, 1983

Standards Australia "Steel Structures", AS 4100 - 1990, 1990

D.M.Thompson, "Loading Tests on Highway Bridges: A Review", Transport and Road Research Laboratory, Dept. of the Environment, Dept. of Transport, TRRL Laboratory Report 1032, 1981

Yen,B.T., Huang,T., Lai,L-Y., Fisher,J.W. “Manual for Inspecting Bridges for Fatigue Damage Conditions”, *Research Report No.85-02, Commonwealth of Pennsylvania Department of Transportation,* Department of Civil Engineering, Lehigh University, Jan.1990

Non-Destructive Inspection-Based Methods for Fatigue Reliability Evaluation

Sankaran Mahadevan[1] and Ruoxue Zhang[2]

Abstract

This paper develops a comprehensive approach to integrate computational reliability methods and non-destructive inspection (NDI) for fatigue reliability evaluation. Uncertainties existing in the performance and outcome of NDI are considered through numerical statistical measures and the mathematical relationships among these measures are derived. A new method is proposed to estimate the probability of detection (POD) and confidence interval using these relationships. Also, a new method of using the NDI results to update the reliability estimation is presented, which includes information such as POD, measurement accuracy, false call probability, and the probability of non-detection of large flaws. The application of the proposed methods is illustrated through an example fatigue problem. Inspection techniques with different POD characteristics are compared.

Introduction

In recent years, much attention is being paid to the deterioration of the nation's infrastructure. One of the important mechanisms of deterioration is the fatigue effect on structures subjected to repeated or cyclic loading. The design fatigue reliability can be estimated by combining reliability analysis methods with fatigue life or fracture mechanics models. Quantitative probabilistic life prediction can be used either during the design stage to account for degradation in a new structure, or to obtain estimates of the adequacy of an existing structure based on projections of the future loads. On the other hand, the nature of the fatigue process and uncertainties associated with the estimation of load histories also require field inspection as a necessary tool for damage detection and prevention. The development of modern

[1] Professor, Dept. of Civil and Environmental Engrg., Vanderbilt Univ., Nashville, TN 37235

[2] Research Assoc., Dept. of Civil and Environmental Engrg., Vanderbilt Univ., Nashville, TN 37235

nondestructive inspection techniques (NDI) may facilitate early detection of flaws and allow a more economic design and help to ensure the safe condition and extend the service life of the structure.

Thus, probabilistic life prediction and regular inspection are two complementary ways for ensuring the safe performance of aging structures, and these two activities can be integrated together. On the one hand, quantitative probabilistic life-prediction can be used for defining suitable inspection intervals and mandating repairs. On the other hand, the results of damage inspection offer information on the state of the structure, and the original life prediction model with many uncertain parameters can be modified based on this newly available information and give a more accurate prediction for the remaining life. Furthermore, this updated model for the probabilistic prediction of remaining life can be used to decide the interval for the next inspection or whether a repair is required immediately. Thus the optimum strategy for reassessment and maintenance of a structure against fatigue is to combine quantitative reliability estimation and NDI. This paper develops a methodology to facilitate this strategy and demonstrates its application to civil structures.

While NDI is an essential tool for the assessment of an aging structure, it introduces additional uncertainty in the reliability analysis due to the uncertainties in the inspection technique itself. Several measures such as probability of detection (POD), flaw size measurement accuracy, and false call probability (FCP) have been used to quantify the uncertainty and reliability of NDI techniques. These criteria describe different aspects of the capability and performance of a given NDI technique. In the NDI profession, they are usually treated as independent quantities and obtained from separate experiments designed to measure each quantity. In this paper, the inherent mathematical relationships among these measures are investigated with a probabilistic framework. It is shown that the use of these relationships will significantly reduce the number of experiments required to characterize the reliability of inspection.

Several studies have been reported to incorporate the information from inspection to update fatigue reliability. Tang (1973) developed a method to use the POD curve of NDI to update the distribution of crack size at the time of inspection, given that crack or no crack is detected. With this updated distribution of crack size at the time of inspection taken as the new initial crack size, the fatigue failure probability with respect to time is reassessed by Byers (1997) and Zheng and Ellingwood (1998). In their work, the case of crack detected with size measurement is also considered. However, only the initial crack size in the crack growth model is updated after inspection, while all the other uncertain parameters and the crack growth model are assumed to remain the same. These assumptions are relaxed in the present study. In another updating technique developed by Madsen (1987), and Zhao and Haldar (1994), different limit state functions are defined corresponding to different inspection results, and the failure probability is updated directly based on inspection results. For the convenience of defining the limit state for different inspection results, instead of the POD function, another NDI performance index called detectability is used. Detectability is defined as the critical crack size for a given NDI technique below which a crack cannot be detected. Since this data is not generally available from NDI reliability measurement, it is either defined as the flaw size with a 50% probability that it will be detected, or taken as a random variable with POD as its cumulative distributive function (Zhao and Haldar, 1994). This method has been extended to NDI-based corrosion fatigue reliability updating by Zhang and Mahadevan (2000). However, compared to POD, detectability is difficult to quantify in practice and provides less information on the ability of the NDI technique in detecting a flaw.

Therefore, a practical and mathematically consistent approach is developed in this paper to combine the information about various NDI performance measures to update the failure probability. The proposed method is mathematically consistent because the same mathematical relationships between various performance measures are applied to all outcomes of NDI. It is shown that the reliability updating techniques proposed in earlier studies for different outcomes are special cases of this general formulation. The proposed methodology is comprehensive in that it includes the possibility of false calls and non-detection of large flaws. It uses commonly available practical measures of NDI performance and does not need the definition of any new measure.

Reliability characterization of NDI technique

This section develops the mathematical relationships between various performance measures of NDI reliability. There are many available NDI methods for assessing the integrity of structural components. Common techniques include visual inspection, magnetic testing (MT), eddy current, ultrasonic testing (UT) and acoustic emission (AE) (Simonen, 1995). The method selected for a given application is based on the type of degradation that is to be detected. For the detection of crack defects in metallic components, ultrasonic testing (UT) is perhaps the most commonly used technique and several studies have been reported to quantify the reliability of UT methods (Cowfer 1989, Dau 1983, Heasler 1990, etc.).

The objective of an NDI system is to enhance the reliability of structural components by detecting flaws that can degrade the strength of the component. Having detected flaws that are of structural significance, corrective actions can then be taken to prevent service failures. Ideally an inspection should detect all flaws and also correctly characterize their size and location, so that only those flaws of structural significance will be repaired. In addition, the ideal NDI system should never incorrectly indicate the presence of flaws that do not actually exist. Nor will the ideal system incorrectly characterize flaws so that significant flaws are left unrepaired or benign flaws are unnecessarily repaired.

However, due to many uncertain factors such as the condition of the structure under inspection and its service environment, the sensitivity of inspection equipment, material imperfections and operator skills, no in-service inspection can be perfect. Probabilistic analysis should be applied to account for the effects of those uncertainties and to evaluate the reliability of performance of a given NDI system. A number of measures are commonly used in the profession to quantify the reliability of an NDI system.

Probability of Detection

The probability of detection (POD(a)) expresses the probability of detecting a flaw of a given size a and is the common measure to evaluate the capability of an NDI technique. The POD curve for a particular NDI method can be obtained by introducing flaws of various sizes into test specimens and by calculating the ratio of number of flaws detected to the number of flaws actually existing. In this approach, a large number of tests need to be done to obtain a POD curve. Generally, POD increases with flaw size and eventually attains a maximum value at which non-detection is governed by other complicating factors (e.g., human errors) that may dominate the detection process (Simonen, 1995). Fig. 1 shows some typical forms of a POD curve. Curve 1 incorporates the possibility that a very large flaw may not be detected, and Curve 2 considers false call probability (i.e., non-zero probability of detection when there is actually no flaw). These two cases will be considered in detail in the following sections. Curve 3, which has been used in some studies, ignores the possibility of false calls and non-detection of large flaws.

Many studies have been reported to develop the POD for different methods. Several POD models in the form of a cumulative distributive function (CDF) of flaw size a have been suggested, including exponential distribution (Tsai and Wu, 1993), lognormal distribution and Weibull distribution (Tang 1973).

As stated before, due to various reasons, even very large flaws may not be detected with certainty, i.e., POD of very large a is not necessarily equal to 1. This fact leads to another expression for POD which is based on the exponential distribution (Staat, 1993):

$$POD(a) = (1-p)[1-\exp(-ca)] \tag{1}$$

where c is a parameter derived from experimental data. This POD(a) is asymptotically equal to 1-p for large values of a. Typically, p would be of the order of 0.01 to 0.05 for flaw sizes of practical interest. The POD curve for this case is illustrated in Fig. 1 by Curve 1.

Size Measurement Accuracy

Size measurement accuracy measures the difference between the true size of a flaw and its estimated size as established by the NDI technique. Measurement errors can be either systematic (i.e., the measured size is consistently larger or smaller than the actual size) or random in nature, and are typically made up of a combination of systematic and random errors. The relationship between the actual and measured flaw size is described by some studies (Rummel et al 1989, Jiao 1989) with a linear function determined by regression analysis, as illustrated in Fig. 2:

$$A = b_0 + b_1 A_m + \varepsilon' \tag{2}$$

where A is the actual flaw size, A_m is the measured flaw size, b_0 and b_1 are regression coefficients, and ε' is the residual random error with zero mean value. In many practical laboratory test programs, actually the true size A is given and A_m is regressed on A, based on the following relationship (Healser, 1990):

$$A_m = \beta_0 + \beta_1 A + \varepsilon \tag{3}$$

ε is generally assumed to be normally distributed with a standard deviation σ_ε, which varies with different NDI techniques. It is also assumed that ε has a constant variance for different flaw sizes. The smaller the σ_ε is, the better the NDI performs. The extreme case that $\beta_0 = 0$, $\beta_1 = 1$ and $\sigma_\varepsilon = 0$ represents the ideal NDI system, which detects a flaw of any size with exact measurement. Besides this linear size measurement model, other probabilistic models can also be applied to describe the relationship between A_m and A, depending on the actual experiment data of a specific NDI technique (Heasler, 1990).

False Call Probability

False call probability (FCP) is defined as the fraction of times that an unflawed component or structural element will be incorrectly classified as being flawed. Repair or replacement due to a false call would cause unnecessary economic cost and may result in potential damage. Usually, high capability of flaw detection is connected to high probability of false call, so there is a trade-off between detection of flaws and rejection of good components.

FCP can be obtained through performing NDI on unflawed components. Comparing the definition of POD and FCP, it can be seen that FCP is actually the value of POD at flaw size $a = 0$. However, ignoring the possibility that a flaw may be indicated even when the actual flaw size is zero, many currently used POD models in literature do not incorporate the false call probability. To account for the false call probability in the POD model, Heasler et al (1990) proposed a POD model as follows:

$$POD(a) = \{1 + \exp[-(\alpha + \beta a)]\}^{-1} \tag{4}$$

where α and β are parameters regressed from experimental data. Thus, the FCP is

$$FCP = POD(0) = [1 + \exp(-\alpha)]^{-1} \tag{5}$$

This POD model with consideration of FCP is shown in Fig.1 by the dashed line with POD(0) ≠ 0.

Relationship Between POD, FCP and Size Measurement Accuracy

Although probability of detection, false call probability and measurement accuracy describe different aspects of NDI performance and are obtained from different experiments, they are not independent measures. The probability that a flaw with size a is detected may be expressed as:

$$POD(a) = P(A_m > 0 | A = a) \tag{6}$$

where A is the actual flaw size, and A_m is the measured flaw size. Considering the linear model in Eq. (3) as an example, we obtain

$$POD(a) = P(\beta_0 + \beta_1 a + \varepsilon > 0) = P(\varepsilon < \beta_0 + \beta_1 a) \tag{7}$$

In the same way, the false call probability is obtained as:

$$FCP = P(A_m > 0 | A = 0) = P(\beta_0 + \varepsilon > 0) = P(\varepsilon < \beta_0) = POD(0) \tag{8}$$

If for a certain NDI technique, a probabilistic model of measurement accuracy has been established (i.e., the relationship between A_m and A and the statistics of the associated parameters), but the POD curve and FCP are unavailable, then, Eqs. (7) and (8) can be used to obtain the POD curve and FCP.

For illustration, following the discussion in the previous section, assume that random error $\varepsilon \sim N(0, \sigma_\varepsilon)$. Using Eq. (7) and Eq. (8), POD and FCP become:

$$POD(a) = \Phi\left(\frac{\beta_0 + \beta_1 a}{\sigma_\varepsilon}\right) \tag{9}$$

and

$$FCP = \Phi\left(\frac{\beta_0}{\sigma_\varepsilon}\right) \tag{10}$$

Consider the effects of β_0 and σ_ε on the probability of detection and the false call probability in Eqs. (9) and (10). σ_ε is related to the steepness of the POD curve. A smaller value of σ_ε, which corresponds to a smaller variance of measurement accuracy, is associated with a steeper POD curve. A negative β_0 reduces the probability of false calls while decreasing the probability of detecting when a > 0, and a positive β_0 has the opposite effect. An inspection technique with a small negative systematic error β_0 and a small random error σ_ε will have a high capability of detection with low false call probability.

Thus, the inherent relationships between the different measures to evaluate the performance of NDI are revealed. In this study, two assumptions are used in the discussion of the proposed method: (1) linear relationship between the measured flaw size and the actual flaw size, and (2) normal distribution of the measurement error (random error). These assumptions are used only to help illustrate the derivation of the inherent relationship among different measures of inspection's uncertainty and the development of the method. Based on actual inspection data, other more realistic measurement accuracy models and random error models may be adopted, and the relationships between POD, FCP and measurement accuracy can be derived similar to those in Eqs. (6) to (10). However, the inherent relationship and the basic method, which is derived mathematically, will be the same, and the consistency of the proposed method will not be affected.

Estimation of POD and Confidence Bounds

As stated in the previous section, in the currently used approach, the POD curve for a particular NDI method is obtained by introducing flaws of various sizes into test specimens and by calculating the ratio of number of flaws detected to the actual number of flaws. Therefore, a large number of tests need to be done to obtain the POD within a desired confidence interval. For example, assume that specimens with different flaw sizes are tested to obtain the POD curve. For each flaw size a, m hit or miss (meaning that a flaw is either found or not found) tests are conducted, of which k flaws are detected. Thus the estimated POD at a is (Shooman, 1968):

$$P\hat{O}D(a) = \frac{k}{m} \tag{11}$$

Based on the central limit theorem, for a large value of m, the term $\dfrac{k - mp}{\sqrt{mp(1-p)}}$ (where p denotes POD (a)) approximately follows a standard normal distribution. Therefore, the confidence interval width for the estimated POD is obtained to be

$$\frac{\sqrt{\left(2k + z_{\alpha/2}^2\right)^2 - 4 \times \left(m + z_{\alpha/2}^2\right) \cdot k^2/m}}{m + z_{\alpha/2}^2} \tag{12}$$

Thus, if j flaw sizes are needed to establish a POD curve, the total number of tests required is $\sum_1^j m_i$, where m_i is the number of tests corresponding to the ith flaw size.

However, in the proposed approach, when size measurement is available from NDI, the inherent relationship between POD and size measurement shown in Eq. (6) may be used as an alternative way to obtain the POD. The size measurement data contains more information than hit/miss data and, as a result, may require a smaller minimum number of samples to conduct an analysis.

For example, assume that the relationship between the measured size and actual size follows the linear model as indicated in Eq. (3). After n tests, the regression coefficients β_0, β_1 and the standard deviation of random error ε in Eq. (3) are estimated as (Hayter, 1996):

$$\hat{\beta}_1 = \frac{\sum_{i=1}^{n}\left(A_i - \overline{A}\right)\left(A_{mi} - \overline{A_m}\right)}{\sum_{i=1}^{n}\left(A_i - \overline{A}\right)^2} \tag{13}$$

$$\hat{\beta}_0 = \overline{A_m} - \hat{\beta}_1 \overline{A} \tag{14}$$

$$\hat{\sigma}_\varepsilon^2 = \frac{\sum_{i=1}^{n}\left(A_{mi} - \left(\hat{\beta}_0 + \hat{\beta}_1 A_i\right)\right)^2}{n-2} \tag{15}$$

where n is the total number of tests, $\overline{A}$ and $\overline{A_m}$ are the mean values of the true flaw size and measured flaw size respectively. In terms of the inference of the measured size A_m corresponding to an actual flaw size A, and the inference of the standard deviation of the random error, we have

$$\frac{\hat{\sigma}_\varepsilon^2}{\sigma_\varepsilon^2} \sim \frac{\chi_{n-2}^2}{n-2} \tag{16}$$

$$\frac{\left(\hat{\beta}_0 + \hat{\beta}_1 A\right) - \left(\beta_0 + \beta_1 A\right)}{\hat{\sigma}_\varepsilon \sqrt{\frac{1}{n} + \frac{\left(A - \overline{A}\right)}{\sum_{i=1}^{n}\left(A_i - \overline{A}\right)^2}}} \sim t_{n-2} \tag{17}$$

where χ_{n-2}^2 denotes a chi-square random variable with (n-2) degrees of freedom, and t_{n-2} denotes a t-distributed random variable with (n-2) degrees of freedom.

Therefore, the estimation of POD(a) in Eq.(6) using size measurement model is obtained from Eq. (9), in which the parameters are estimated in Eqs. (13) to (15). The confidence interval is estimated from Eqs. (16) and (17).

Numerical Example

A numerical example is considered here to illustrate the estimation procedure and compare the number of tests needed for the two methods to achieve the same confidence interval width. Assume, for the sake of this example, that the true POD of the NDI technique is as shown in Fig. 3, and that the measured flaw size is related to the actual flaw size as in Eq. (3) with β_0 = -1.8, β_1 = 1.2, and σ_ε = 1.5. (In reality, these quantities are not known. One can only estimate these quantities from either hit/miss tests or size-measurement tests. The above data is assumed only to compare the two approaches on the same basis). The above assumed data is used with Monte Carlo simulation in this example to generate "observations" for the two approaches. For the proposed size measurement-based approach, 100 different flaw sizes are chosen. For each flaw size, a value of the random error ε is sampled with a normal distribution (zero mean and σ_ε = 1.5). Then Eq. (3) gives the "measured" flaw size. Then, using regression analysis on these 100 "observations", the coefficients of Eq. (3) are estimated as $\hat{\beta}_0$ = -1.97, $\hat{\beta}_1$ = 1.24, and $\hat{\sigma}_\varepsilon$ = 1.441. (Note that as the number of observations increases, the estimates will converge to the true values). Next, Eq.

(9) is used to estimate the POD and the corresponding confidence interval for the size measurement-based approach. The POD curve obtained using Eq. (9) is shown in Fig. 3.

Fig. 4 shows the confidence interval width of the POD estimated using Eqs. (16) and (17). The number of hit/miss tests needed to achieve the same confidence interval width corresponding to different flaw sizes is calculated using Eq. (12) and illustrated in Fig. 5. In solving Eq. (12), a Monte Carlo-based semi-inverse approach is used. For example, start with an initial guess of m = 100 for a particular flaw size a_i. 100 values of the POD are simulated with a uniform (0,1) distribution, and k is the number of samples with POD value lower than the true value. These values of k and m are substituted in Eq. (12) to check if the confidence interval width is the same as that from the proposed approach. If not, a different value of m is assumed, and the procedure is repeated, until convergence is achieved. The value of m that corresponds to the converged solution of Eq. (12) is reported in Fig. 5, and the corresponding k/m value is reported as the POD estimate from the hit/miss approach in Fig. 3.

Assume that specimens with 11 different flaw sizes, uniformly distributed between 0 and 5 mm, are tested in order to draw a POD curve. A total of $\sum m_i$ = 6171 tests are needed in the hit/miss approach to get the same confidence interval width as the proposed size measurement-based approach, obtained with only 100 tests. If 5 different flaw sizes (A = 0, 1, 2, 3, 4 mm) are inspected, a total of $\sum m_i$ = 1368 are needed. Therefore, if the size measurement-based method is used to estimate the POD curve, the amount of tests may be significantly reduced.

Fatigue Reliability Prediction

The S-N curve-based approach and linear-elastic fracture mechanics (LEFM) approach are two common methods used for fatigue life estimation and fatigue reliability analysis. Since LEFM method explicitly considers crack size and can be incorporated with inspection results to update the fatigue-damage evaluation, it is adopted in this paper as the approach for fatigue reliability estimation.

Several empirical models are available to describe the fatigue crack growth rate, e.g., Paris law (Paris, 1964), Foreman's equation (Foreman, et al, 1967) and Weertman's Equation (Weertman, 1966). Paris law is the simplest and most widely used crack growth model, and is used herein:

$$\frac{da}{dN} = C(\Delta K)^m \tag{18}$$

where a is the crack size, N is the number of stress cycles, C and m are fatigue crack growth parameters (material constants that are determined experimentally), and ΔK is the stress intensity range:

$$\Delta K = K_{\max} - K_{\min} = F(a, \mathbf{Y})\sqrt{\pi a}\Delta\sigma \tag{19}$$

where σ is the far-field stress, $\Delta\sigma = (\sigma_{\max} - \sigma_{\min})$, $F(\alpha, \mathbf{Y})$ is the geometric shape function to account for the possible stress concentration, and $\mathbf{Y}$ is the vector of random variables affecting the shape function, such as the stress concentration coefficient and the dimensions of the specimen under consideration. Equation (18) can be integrated to estimate the fatigue life. The number of cycles required to propagate a crack from an initial size, a_0, to size a is then given by:

$$N = \int_{a_0}^{a} \frac{da}{C(\Delta K)^m} \tag{20}$$

The fatigue failure criterion for a structure subjected to N stress cycles is defined as:

$$a_c - a_N \leq 0 \tag{21}$$

where a_N is the in-service crack size corresponding to N stress cycles and can be obtained by Eq. (20), and a_c is the critical crack size. a_c can be defined as the size of the crack causing failure or a design crack size beyond which the serviceability requirements cannot be satisfied. Eq. (21) is adopted as the limit state function of the fatigue reliability problem considered herein. The corresponding failure probability is then

$$P_f = P[g(\mathbf{Z}) \leq 0] = P[a_c - a_N \leq 0] \tag{22}$$

where $\mathbf{Z}$ denotes the group of all the basic random variables in the crack growth model, including C, m, a_0, and $\mathbf{Y}$. To perform reliability analysis, the uncertainties associated with these random variables must first be defined and quantified. Then the first-order reliability method (FORM), the second-order reliability method (SORM) or Monte Carlo simulation may be used to estimate the failure probability in Eq. (22).

Fatigue Reliability Updating

The analytical fatigue reliability estimation in the previous section provides a forecast of the probabilistic time-dependent behavior of the structural component, while regular nondestructive inspection gives additional information on the in-service condition. Since there are uncertainties in the analytical estimation and also uncertainties in the inspection process, as discussed and quantified earlier, the analytical prediction, inspection data and uncertainties should be integrated to update the fatigue reliability estimate and to make decisions for subsequent maintenance or repair.

Three possible outcomes of inspection are considered: no crack detected, crack detected without size measurement and crack detected with size measured. The event in the third case has been defined through a limit state function by Madsen (1987, 1997), and Zhao and Haldar (1994) as:

$$a_{N_0} - A = 0 \tag{23}$$

where N_0 is the number of stress cycles the component has experienced at the time of inspection and A is the crack size at the time of inspection. Considering the uncertainty in the inspection process, the crack size A is not a deterministic value equal to the measured result but a random variable. Its statistical property can be obtained through the measurement accuracy model as discussed in the previous section. The updated failure probability in this case has been obtained by Madsen (1987, 1997) as follows:

$$\begin{aligned} P_{f,up} &= P[g(Z) \le 0 | a_{N_0} - A = 0] \\ &= \frac{\frac{\partial}{\partial x} P\left[a_c - a_N \le 0 \le 0 \cap a_{N_0} - A \le x\right]_{x=0}}{\frac{\partial}{\partial x} P\left[a_{N_0} - A \le x\right]_{x=0}} = \Phi\left(-\beta_{up}\right) \end{aligned} \tag{24}$$

where the derivatives are computed for $x = 0$. An evaluation of the sensitivity factor for a parallel system (numerator) and a component (denominator) are thus required, and a FORM or SORM analysis can be directly applied. The derivative calculation can also be performed approximately through calculating the difference of probability at two very small x values around 0, using FORM, SORM or Monte Carlo simulation.

However, in the case that no crack is detected or crack is detected but with no size measurement, the updating procedure is not so straightforward. In order to apply the same procedure as in Eq. (24), a limit state must be defined for these two cases. Madsen (1987, 1997), Zhao and Haldar (1994) adopted the following limit state:

For $\bar{D}$: no crack detection: $\quad a_{N_0} - a_d < 0 \qquad$ (25a)

For D: crack detection $\quad a_{N_0} - a_d > 0 \qquad$ (25b)

where a_d is defined as the detectability of the NDI technique, i.e., the detectable crack size, and is taken as an deterministic or random variable. To make Eq. (25) applicable, a_d is taken in these studies as the flaw size with a 50% probability that it will be detected or a random variable with its cumulative distribution function (CDF) equal to the POD function. The first definition reflects the capability of the NDI system to some degree, but obviously loses much information provided by the POD curve which is often available in the profession, including the steepness of the curve which represents the random error of the technique and the false call probability. The second definition can be applied only when the POD function satisfies the definition of a CDF. This is the case with Curve 3 in Fig. 1. However, for the case that the false call probability is not equal to 0, as in Eq. (4), or when there is a probability that even very large flaws may not be detected, as in Eq. (1), the POD curve cannot be assumed to be the CDF of a_d.

Thus, the POD function should be incorporated directly into the failure probability updating to account for the uncertainties in the NDI performance more comprehensively, including false call probability. Actually, the POD function has been used by Tang (1973), Byers (1997), Zheng and Ellingwood (1998) to update the distribution of crack size at the time of inspection, given that crack or no crack is detected. However, in these studies, the instant of this inspection is taken as the new initial time and the updated distribution of crack size at this time taken as the new initial crack size. Then the fatigue failure probability in the remaining life is reassessed, using the same crack growth model in Eqs. (20) and (21). Therefore, in this procedure, actually only the initial crack size in the crack growth model is updated, while all the other uncertain parameters associated with the crack growth model remain the same. In the context of reliability updating, this is not a full Bayesian way. Therefore, in this section, a new comprehensive approach is developed to incorporate the outcome and uncertainties of inspection to update the failure probability.

First, the case of no crack detection is considered. The event is denoted by $\overline{D}$. Using the Bayes theorem, the updated failure probability is the following conditional probability:

$$P_{f,up} = P\left(a_c - a_N \le 0 \middle| \overline{D}\right) = \frac{P\left(\overline{D} \middle| a_c - a_N \le 0\right) P\left(a_c - a_N \le 0\right)}{P\left(\overline{D}\right)} \tag{26}$$

In this equation,

$$P\left(\overline{D}\right) = \int_0^\infty \left(1 - POD(a)\right) f_{a_{No}}(a)\,da \tag{27}$$

where $f_{a_{No}}(a)$ is the probability density function (PDF) of the crack size at the time of inspection, and is the derivative of the CDF of crack size, $F_{a_{No}}(a)$. $F_{a_{No}}(a)$ may be obtained directly from reliability estimation at stress cycle N_0 as follows:

$$F_{a_{No}}(a) = P\left(a_{N_0} - a \le 0\right) \tag{28}$$

where a_{N0} is obtained using the probabilistic crack growth model of Eq. (20).

With the same theorem, the first term of the numerator in Eq. (26) is:

$$P\left(\overline{D} \middle| a_c - a_N \le 0\right) = \int_0^\infty \left(1 - POD(a)\right) f_{a_{No}}(a | a_c - a_N \le 0)\,da \tag{29}$$

where $f_{a_{No}}(a | a_c - a_N \le 0)$, the conditional PDF of crack size, may be obtained as:

$$f_{a_{No}}(a | a_c - a_N \le 0) = \frac{dF_{a_{No}}(a | a_c - a_N \le 0)}{da} = \frac{dP(a_{No} \le a | a_c - a_N \le 0)}{da} \tag{30}$$

Substituting Eq. (27) through Eq. (30) into Eq. (26), we obtain

$$P_{f,up}=\frac{\left[\int_0^\infty (1-POD(a))\frac{dP(a_{No}\le a|a_c-a_N\le 0)}{da}da\right]P(a_c-a_N\le 0)}{\int_0^\infty (1-POD(a))\frac{dP(a_{No}\le a)}{da}da}$$
$$=\frac{\left[(1-POD(\infty))+\int_0^\infty P(a_{No}\le a|a_c-a_N\le 0)\frac{dPOD(a)}{da}da\right]\times P(a_c-a_N\le 0)}{(1-POD(\infty))+\int_0^\infty P(a_{No}\le a)\frac{dPOD(a)}{da}da}$$
$$=\frac{(1-POD(\infty))P_f+\int_0^\infty P(a_{No}\le a\cap a_c-a_N\le 0)\frac{dPOD(a)}{da}da}{(1-POD(\infty))+\int_0^\infty P(a_{No}\le a)\frac{dPOD(a)}{da}da}$$
(31)

In this approach of failure probability updating, the definition of a limit state function of crack detection as in Eq. (25) is avoided while all the available information on inspection results and inspection performance is incorporated.

The same approach can be applied to the case that a crack is detected but with no size information, which is denoted by event D as follows:

$$P_{f,up}=P(a_c-a_N\le 0|D)=\frac{P(D|a_c-a_N\le 0)P(a_c-a_N\le 0)}{P(D)}$$
$$=\frac{\int_0^\infty POD(a)\frac{dP(a_{No}\le a|a_c-a_N\le 0)}{da}da\cdot P(a_c-a_N\le 0)}{\int_0^\infty POD(a)\frac{dP(a_{No}\le a)}{da}da} \quad (32)$$
$$=\frac{POD(\infty)\times P_f-\int_0^\infty P(a_{No}\le a\cap a_c-a_N\le 0)\frac{dPOD(a)}{da}da}{POD(\infty)-\int_0^\infty P(a_{No}\le a)\frac{dPOD(a)}{da}da}$$

In Eqs. (31) and (32), the term dPOD(a)/da is obtained directly from the experiment-based POD function. $P(a_{No}\le a\cap a_c-a_N\le 0)$ may be calculated using Monte Carlo simulation or a FORM-based analysis for a parallel system.

As a special case, assume that FCP = 0 and that a very large flaw can be detected with certainty (which means POD(∞) =1). In this case, the POD function satisfies the criteria for a valid CDF, and may be assumed to be the CDF of the detectability a_d. Therefore, Eq. (31) becomes:

$$
\begin{aligned}
P_{f,up} &= \frac{\int_0^{\infty} P(a_{No} \le a_d \cap a_c - a_N \le 0 | a_d = a) f_{a_d}(a) da}{\int_0^{\infty} P(a_{No} \le a_d | a_d = a) f_{a_d}(a) da} \\
&= \frac{P(a_{No} \le a_d \cap a_c - a_N \le 0)}{P(a_{No} \le a_d)}
\end{aligned}
\tag{33}
$$

This formula is the same as the one derived by Zhao and Haldar (1994) by defining the limit state in Eq. (17*a*). Similarly, the updated failure probability with flaw detection is obtained from Eq. (32) as:

$$
\begin{aligned}
P_{f,up} &= \frac{P_f - \int_0^{\infty} (a_{No} \le a_d \cap a_c - a_N \le 0 | a_d = a) f_{a_d}(a) da}{1 - \int_0^{\infty} P(a_{No} \le a_d | a_d = a) f_{a_d}(a) da} \\
&= \frac{P(a_c - a_N \le 0) - P(a_{No} \le a_d \cap a_c - a_N \le 0)}{1 - P(a_{No} \le a_d)} \\
&= \frac{P(a_{No} \ge a_d \cap a_c - a_N \le 0)}{P(a_{No} \ge a_d)}
\end{aligned}
\tag{34}
$$

which is the same as the result derived by Zhao and Haldar by defining the limit state in Eq. (25b).

Therefore, the concept of detectability can be applied and taken as a random variable with the POD function as its CDF, as a special case of the general formulas in Eqs. (31) and (32), which incorporate all the information provided by the POD curves normally available in the profession.

Numerical Example

Two full-penetration butt welds in the tension flange of a simply supported steel bridge are considered. The fatigue reliability of these full-penetration butt welds is considered. It is assumed that the butt welds are subjected to a train loading of 100 stress cycles per day. This represents approximately 1,825,000 stress cycles for a design life of 50 yr. The width of the tension flange is 1.0 m.

To perform the reliability analysis, the uncertainty of the basic random variables in the limit state function associated with the two crack models must first be quantified. Since the crack growth rate is very high near the critical crack size, the effect of the critical size a_c on the fatigue failure is rather small compared to other random variables involved in the entire fatigue damage process. Thus, a_c may be assumed to be a deterministic parameter. For illustrative purposes, the critical crack size is considered to be 0.02 m in this example.

The distributions most frequently suggested in modeling the uncertainty in the initial crack size are the lognormal and the Weibull distributions. The lognormal distribution tends to be more conservative since it predicts a higher probability of large cracks. Agerskov (1999) indicated 0.075 mm and 0.4 mm as the lower and upper bounds of initial crack size of welded joints of steel bridges. In this paper, the initial crack size a_0 is modeled by a lognormal variable with a mean value of 0.2 mm and a coefficient of variation of 0.50.

As to the shape function, it is generally recognized to be a function of the crack size and a group of other random variables such as the stress concentration coefficient and the dimension of the specimen (Fisher, 1984). Closed-form models to express the shape functions are available only for special cases. The geometry function proposed by Paris (1964) to treat the case of center-notched specimens of bridges is used in this study as an illustration:

$$F(\xi)=\frac{1-0.5\xi+0.37\xi^2-0.044\xi^3}{\sqrt{1-\xi}} \tag{35}$$

where $\xi = 2a/w$ and w = width of specimen (in this case, it's the width of the tension flange, 1.0 m).

It is observed that for bridge structures, the Rayleigh distribution would be the most appropriate distribution for the stress range $\Delta\sigma$ (Schilling et al, 1974):

$$f(\Delta\sigma)=\frac{\Delta\sigma}{Var}\exp\left(-\frac{1}{2}\left(\frac{\Delta\sigma}{Var}\right)^2\right) \tag{36}$$

where *Var* is the variance of stress. In this paper, *Var* is assumed to be 900 MPa^2.

For the material properties C and m, it is observed that for structural steel, the fatigue growth parameter C approximately follows a lognormal distribution while the fatigue growth exponent m is assumed to be a deterministic variable (Zhao and Haldar, 1994). Based on the literature (Fisher, 1984), the mean value of C is taken to be 3.82×10^{-12} using units of meters for crack size and MPa$\sqrt{\text{m}}$ for ΔK. The coefficient of variation (COV) of C is assumed to be 0.6, and m is assumed to be 3.0 in this study.

The reliability index of the butt weld is then evaluated using Eq. (21) and shown in Fig. 6.

Now suppose that the full-penetration butt weld is inspected at about N_0 = 500,000 and no crack is detected. Two POD models for detection techniques, one defined by Eq. (1), which accounts for the possibility that very large flaws may not be detected with certainty, and the other defined by Eq. (4), which incorporates the false call probability, are adopted in this paper. To investigate the effects of the capabilities of different NDI methods, different parameters are used for the two POD models and are illustrated in Fig. 7(a) and Fig. 7(b). Fig. 7(a) incorporates different probabilities that very large flaw may not be detected, which is denoted by p in Eq. (1), and Fig. 7(b) accounts for the false call probability. In both cases, different amounts of steepness of the POD curves are considered.

The updated reliability indices after the inspection with no crack detection are shown in Fig. 8(a) and Fig. 8(b). In all cases, the posterior reliabilities are higher than the prior one, because no crack is detected. It can be seen that the updated reliability index varies with different values of p, FCP and steepness. A smaller p value and a steeper POD curve result in a higher updated reliability index, which means that the performance quality of the NDI technique is incorporated in the reliability updating. However, as the number of cycles increases, the differences between the updated results from various inspection techniques diminish and are in fact closer to the prior computational prediction before inspection. The better the technique, the slower the convergence towards prior computational prediction.

The inspection result of crack detection with no size measurement is also considered herein. In this case, the updated reliability immediately after inspection is shown in Table 1. Curves similar to Fig. 8 may also be developed to show the variation of reliability index with time. However, in this example, considering the low values of the reliability index immediately after inspection, it would be more appropriate to re-inspect or repair the structure. It is observed that crack detection by poorer inspection performance results in more reduction in the estimated reliability.

The updated reliability in the case of crack detection with size measurement (Eq. (24)) has already been studied and illustrated through numerical examples by Madsen (1987) and Zhao and Haldar (1994). As mentioned earlier, the methods used by these studies are special cases of the general method proposed in this paper. The proposed method obtains the same results as in these studies.

Note that the formulation and numerical example in this paper do not consider model uncertainty arising from several sources. The Paris law is used to model crack growth, while other models by Foreman, Weertman etc. are available. The lognormal distribution is used to model the uncertainty in initial crack size, whereas the Weibull distribution is also appropriate. Thus, there is model uncertainty due to competing models. Zhang and Mahadevan (2000) have developed a Bayesian procedure to include this type of model uncertainty in NDI-based fatigue reliability updating. Initially, the predictions from the competing models are combined through a weighted sum, and the weights are updated based on inspection results.

Conclusion

This paper has developed a consistent mathematical basis for the reassessment of time-dependent reliability using non-destructive inspection. The relationships between different measures that characterize the uncertainties in inspection are clearly developed. The use of these relationships, with flaw size measurement, is shown to result in significantly fewer tests for estimating the POD curve and confidence bounds. The NDI performance measures are integrated with computational prediction for reliability re-evaluation. The reliability updating in this study is mainly focussed on the application of the probability of detection (POD) curve, which is commonly used in the profession. Using the proposed updating method, all the features and information provided by the POD curve, including the false call probability and the probability that a large flaw may not be detected, can be addressed. It is also clarified that the methods used in earlier studies, which define limit state functions corresponding to different detection results through the concept of detectability, are special cases of the general method developed in this study. The numerical example demonstrates that the proposed method provides reliability updates that are consistent with the uncertainty measures of the NDI technique.

Acknowledgement

The study was supported by the National Science Foundation under grant no. 9872342 (Program Director: Dr. S.C. Liu). The support is gratefully acknowledged.

References:

Agerskov, H., and Nielsen, J.A. (1999). "Fatigue in steel highway bridges under random loading." *Journal of Struct. Engrg.* ASCE, 125(2), 152-162.

Byers, W.G., Marley, M.J., Mohammadi, J., Nielsen, R.J., and Sarkani, S. (1997). "Fatigue reliability reassessment procedures: state-of-the art paper." *Journal of Struct. Engrg.* ASCE, 123(3), 271-276.

Cowfer, D. (1989). "Basis/back ground for ASME code. Section XI. Proposed Appendix VIII: Ultrasonic examination performance demonstration." *Nondestructive Evaluation: NDE Planning and Application.* ASME, New York.

Dau, G.J. (1983). *Ultrasonic Sizing Capacity of IGSCC and its Relation to Flaw Evaluation Procedures*, Electric Power Research Institute, Charlotte, North Carolina.

Fisher, J. W. (1984). *Fatigue and Fracture in Steel Bridges.* John Wiley & Sons, New York.

Foreman, R.G., Keart, V.E., Engle, R.M. (1967). "Numerical analysis of crack propagation in cycle-loaded structures." *Journal of Basic Engineering*, ASME, 89, 459-464.

Hayter, A.J. (1996). *Probabilistic and Statistics*. Electric Power Research Institute, Boston, MA.

Heasler, P. G., Taylor, T.T. and Doctor, S.R. (1990). "Statistically based reevaluation of PISC-II round robin test data." NUREG/CR-5410, US nuclear Regulatory Commission, Washington, DC.

Jiao, G.Y. (1989). "Reliability analysis of crack growth under random loading considering model updating." PhD thesis, Norweigian Institute of Technology, Trondheim, Norway.

Madsen, H.O. (1987). "Model updating in reliability theory." *Proceeding of ICASP-5*, Vancouver, 564-577.

Madsen, H.O. (1997). "Stochastic modeling of fatigue crack growth and inspection." *Probabilistic Methods for Structural Design*. C. Guedes Soares, ed., Kluwer Academic Publishers, Netherlands, 59-84.

Rummel, W.D., Hardy, G.L., and Cooper, T.D. (1989). *Metals Handbook*, Vol. 17.9th ed. ASM International.

Paris, P.C. (1964). "The fracture mechanics approach to fatigue." *Fatigue-an Interdisciplinary Approach,* Syracuse University Press, New York, 107-132.

Schilling, C. G., et al. (1974). "Fatigue of welded steel bridge members under variable amplitude loads." *Res. Result Digest. No. 60*, Highway Research Board, Washington, D.C..

Shooman, M.L. (1968). *Probabilistic Reliability: An Engineering Approach.* McGraw-Hill, New York.

Simonen, F.A. (1995). "Nondestructive examination reliability." *Probabilistic Structural Mechanics Handbook.* C. Sundararajan, ed., ITP company. New York, 238-260.

Staat, M. (1993). "Sensitivity of and influences on the reliability of an HTR-module primary circuit pressure boundary." *Trans. 12th International Conference on Structural Mechanics in Reactor Tech.*, Elsevier, Amsterdam, 147-152.

Tang, W. H. (1973). "Probabilistic updating of flaw information." *Journal of Testing and Evaluation*, 1(6), 459-467.

Taylor, T. T., Spanner, J. C., Heasler, P.G., Doctor, S. R., and Deffenbaugh, J. D. (1989). "An evaluation of human reliability in ultrasonic inservice inspection for intergranular stress corrosion cracking through round robin testing." *Materials Evaluation*, 47(3), 338-346.

Tsai, C. H. and Wu, W.F. (1993). "Application of probabilistic fracture mechanics to risk assessment of pressure vessels." *Trans. 12th International Conference on Structural Mechanics in Reactor Tech.,* Elsevier, Amsterdam, 135-140.

Weertman, J. (1966). "Rate of growth of fatigue cracks calculated from the theory of infinitesimal dislocations distributed on a plane." *International Journal of Fracture Mechanics*, 2, 460-467.

Zhang, R., and S. Mahadevan, "Model Uncertainty and Bayesian Updating in Reliability-Based Inspection," Structural Safety, Vol. 22, pp. 145-160, 2000.

Zhang, R., and S. Mahadevan, "Reliability-Based Reassessment of Corrosion Fatigue Life," Structural Safety, Vol. 23, pp. 77-91, 2001.

Zhao, Z. Haldar, A., and Breen, Jr.F.L. (1994). "Fatigue-reliability updating through inspection of steel bridges." *Journal of Struct. Engrg*. ASCE, 120(5), 1624-1641.

Zheng, R. and Ellingwood, B. (1998). "Role of non-destructive evaluation in time-dependent reliability analysis." *Structural Safety*, 20(4), 325-339.

Table 1. Updated Reliability at $N = 500{,}000$ with Crack Detected

	Before inspection	After inspection: Eq. (1)			After inspection: Eq. (4)		
		POD 1	POD 2	POD 3	POD 1	POD 2	POD 3
Reliability Index	4.302	-0.327	1.020	1.020	0.215	-0.498	1.242

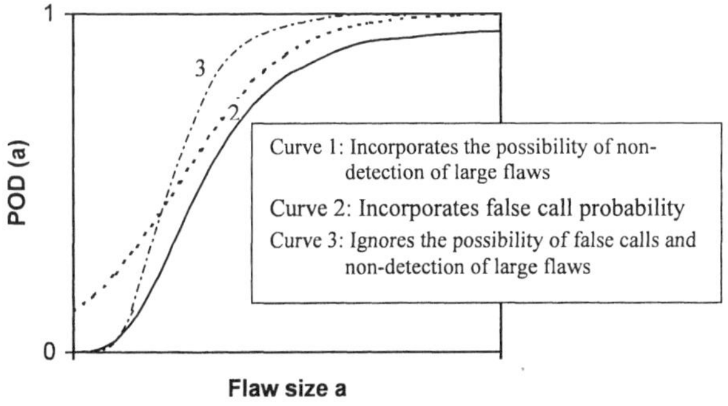

Fig. 1. Illustration of POD Models

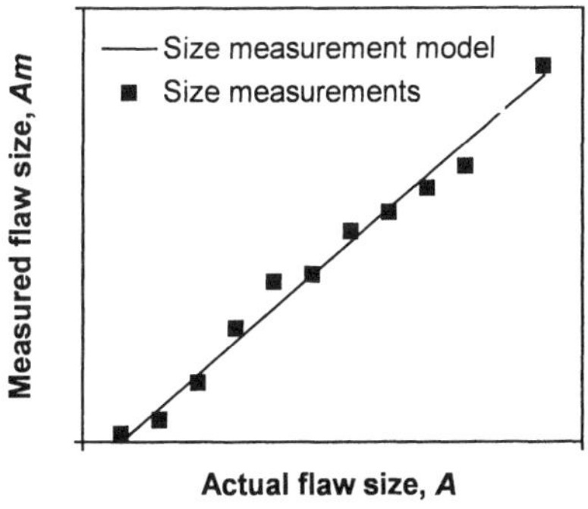

Fig. 2. Regression of Measurement Accuracy Model

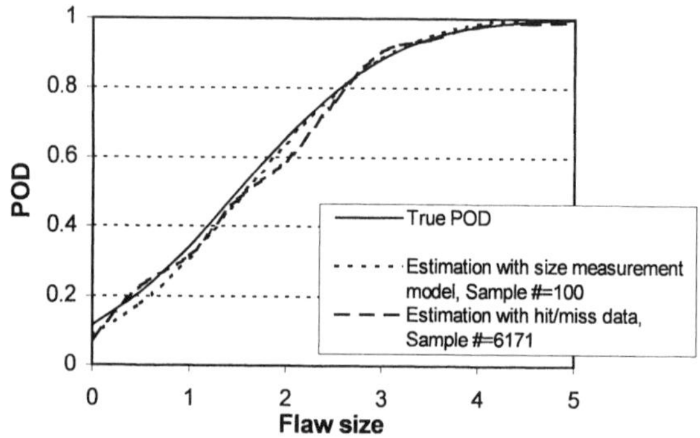

Fig. 3. Estimation of POD Curve

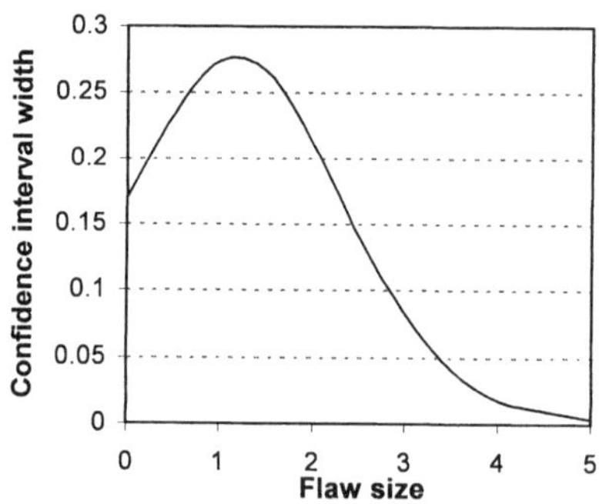

Fig. 4. Confidence Interval of POD Using Size Measurement Model

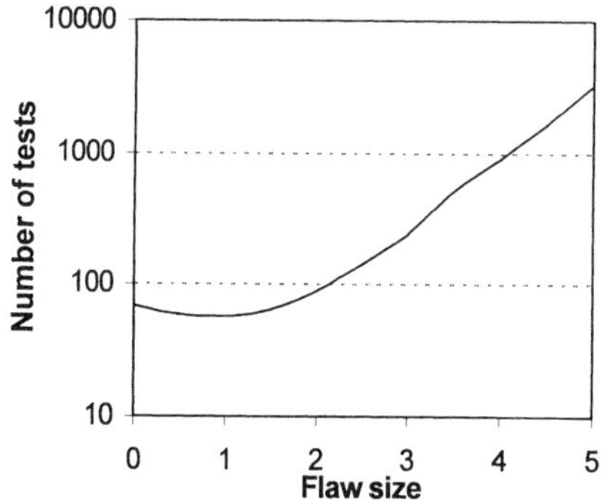

Fig. 5. Number of Hit/Miss Tests for the Same Confidence Interval as in Fig. 4

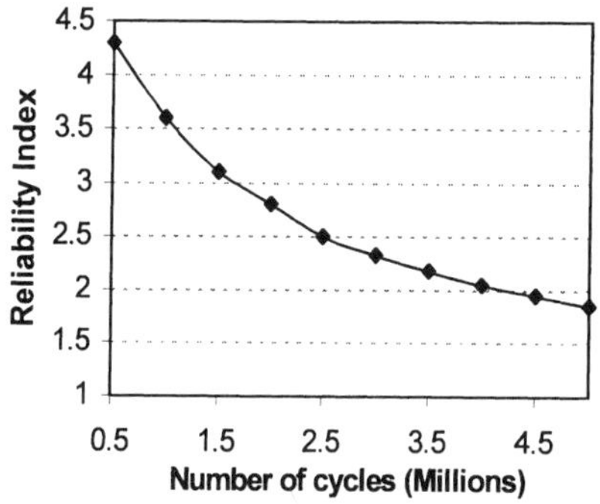

Fig. 6. Reliability Index of the Butt Weld

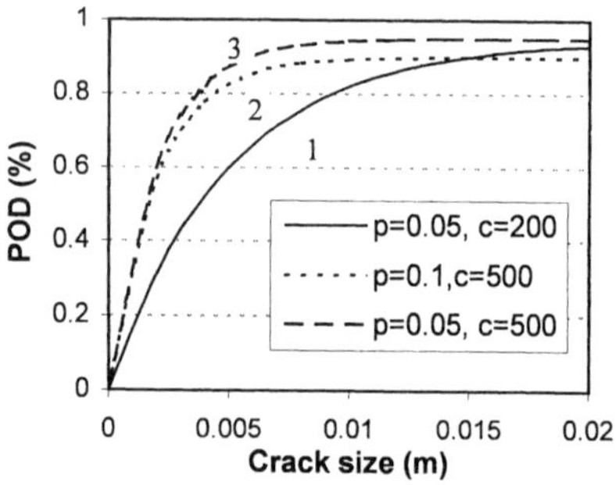

Fig. 7(a). POD Curve of Eq. (1)

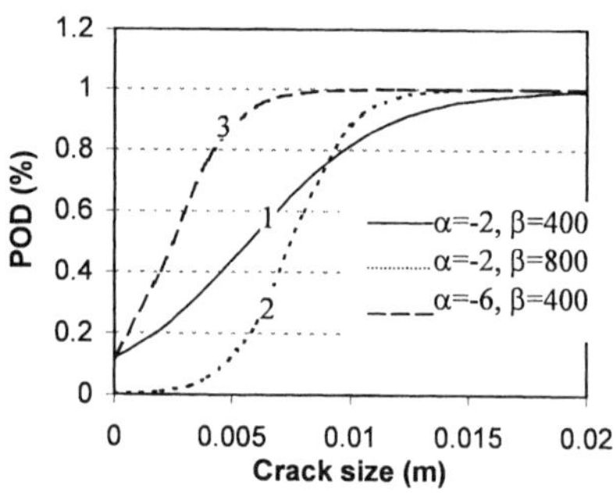

Fig. 7(b). POD Curve of Eq. (4)

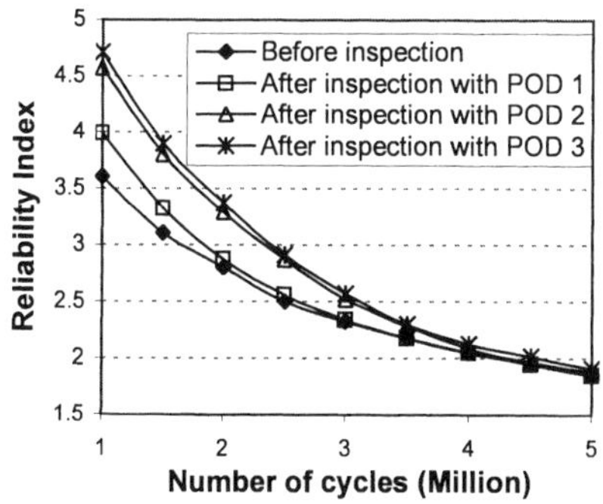

Fig. 8(a). Updated Reliability with No Crack Detected at $N = 500{,}000$ with POD of inspection corresponding to Eq. (1)

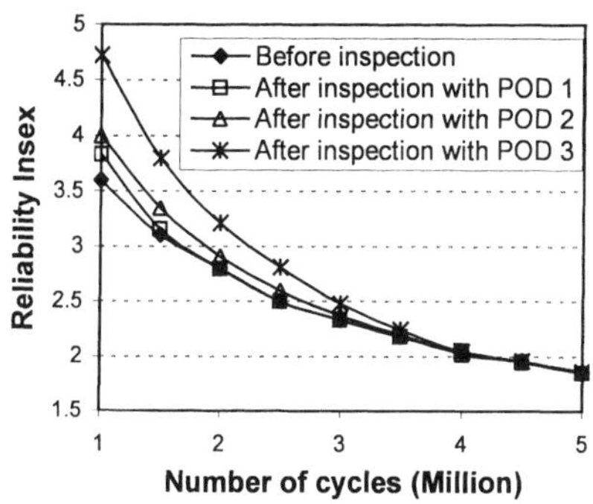

Fig. 8(b). Updated Reliability with No Crack Detected at $N = 500{,}000$ with POD of inspection corresponding to Eq. (4)

Use of Stress Range Data in Fatigue Reliability Assessment of Highway Bridges

J. Mohammadi[1], MASCE, S. A. Guralnick[1], FASCE and R. Polepeddi[2]

Abstract

This paper presents a specific application in which fatigue reliability of steel girders in bridges is assessed by conducting a field measurement of stresses at key critical locations. The procedure presented is regarded as a non-destructive test method. The source of information for fatigue reliability assessment is primarily a set of stress range data collected for fatigue-critical locations in a given bridge. The compiled data is then used along with a rigorous structural analysis to estimate the extent of fatigue damage in components and the assessment of fatigue reliability of the bridge. The data compiled for this study was from several steel girder bridges. The data acquisition session for each bridge lasted for two days. It was believed that within the two-day period a major percentage of potential stress ranges experienced by each bridge would be captured. Following the analysis for the assessment of reliability of these bridges; a study was conducted to determine the cost of bridge replacement based on the extent of fatigue damage in each bridge and the estimates of bridge remaining useful lives. The paper recommends a procedure that can effectively be used in data gathering; and in a follow-up reliability and life cycle cost analysis for highway bridges.

Introduction

The increase in the load and the growth in the truck traffic, occurred in the past several decades, have caused a rapid fatigue damage accumulation in many steel girder bridges. Fatigue damage primarily depends on: (i) the severity of the stress range level; (ii) the number of cycles of the applied stresses; and (iii) the type of structural detail.

---

1 Civil and Arch. Engrg. Dept., Illinois Inst. of Technology, Chicago, IL 60616
2 J.F. Sato and Associates, Littleton, CO 80120

Several categories of fatigue-critical components are specified in the current design guidelines (e.g., AASHTO) and have been utilized in fatigue design and reliability evaluation of structures. In general, the guidelines provide for a safe maximum stress range level based on an estimated fatigue life (i.e., an estimated number of load cycles). For the most part, the margin of safety inherent in the design guidelines is satisfactory. However, bridges usually undergo changes during their service lives wherein the stress history, stress range level, previous damage history, volume of traffic, usage of roadway lanes by vehicles, etc. change; and, as such, fatigue damage may dramatically change also. To account for these factors, more refined procedures for fatigue evaluation of bridges have been proposed (Moses, et al, 1987). To properly include the effect of nearly all factors that affect fatigue life of an existing bridge, it is more appropriate to obtain a realistic set of data on the stress history (which depends on the population of heavy vehicles using the bridge). This can be achieved by conducting a comprehensive program of traffic monitoring and by compiling data on the resulting stress history of the bridge. To avoid the formidable costs associated with a comprehensive and continuous monitoring program, the strain history of a bridge at only a few critical locations may be obtained and used in the fatigue evaluation purposes. The stress range data compiled for these critical locations and the analyses conducted offer an effective non-destructive method for fatigue and fracture reliability assessment of bridges. Similar studies have also been conducted by others utilizing truck load data, such as those obtained through weigh-in-motion (WIM) systems (e.g., Laman and Nowak, 1996).

This paper presents an overview of the process via which fatigue reliability of bridges can be assessed using stress range data from field observations. The application of this method as a non-destructive approach in bridge condition assessment is presented. Furthermore, it is demonstrated that the approach can be used along with a comprehensive life-cycle cost analysis of bridges as a part of a bridge management system.

Bridge Stress Range Data

Field data constitutes a major component of fatigue reliability assessment process. A candidate bridge can be monitored for several days by using strain gages. The installation of gages at critical locations on steel girders of the bridge requires no alteration to the bridge and can be achieved without closing the bridge to the traffic. No additional external load is required; and as such, the bridge will not be subjected to any potential permanent damage or destruction. The strain data compiled can be converted into stress ranges and arranged in a form of frequency diagrams as the data acquisition process is in progress. The development of stress range frequency data may be made by means of a data compression and reduction process, called "cycle counting". Various methods for cycle counting are available. One widely-used method is the "rainflow" process, which is described in standards of the American Society for Testing and Materials (ASTM). A major issue in compiling data to construct stress range frequency is the duration of the data acquisition process. In some applications (e.g., airframe systems), a sufficiently long period of time is needed to capture a realistic representation of all stress ranges and their respective frequencies. In highway bridges, a two or three-day period seems to be satisfactory (Mohammadi, et al, 1993). During this period nearly all stress ranges, experienced by a critical component in a bridge, are captured. Preferably, the data acquisition device should be left running continuously to capture the effect of day and night traffic, and empty and full trucks. To include any seasonal change in the traffic pattern, the data acquisition process may have to be conducted in several two or three-day sessions at various times during the year. It is noted that, usually the compiled data do not contain the stress ranges from "overloaded trucks." These trucks are occasionally allowed to use bridges with special permits that can only be issued by the state transportation authorities.

Figures 1-5 present several samples of stress range data compiled for bridges in Illinois. The example bridges investigated are steel girder structures with reinforced concrete deck slabs. The monitoring system consisted of: (1) an automated data acquisition system; (2) strain gages applied to the bridge girders; and (3) the power supply. Strain gages were installed on the bottom flanges of each bridge's girders at fatigue-critical locations. Nearly all of these bridges have welded cover plates that are considered to be fatigue-critical. Shown in Figs. 1-5 are also theoretical values obtained for the stress ranges. These theoretical values were obtained by means of the Beta probability density function (Ang and Tang, 1975) which was validated for the stress range data of each bridge. The theoretical stress range distribution was used in a probabilistic analysis of fatigue life estimation as explained later in this paper. As seen in Figures 1-5, the maximum stress range values are generally quite low and lie in the 27-34 Mpa (4-5 ksi) range. It is also noted that the stress range data correspond to the live loads on the bridge. The dead load stress is a constant value superimposed on the live load stress range. For a stress time history with relatively constant amplitudes, the mean stress (average of peak and valley stresses) is equal to the dead load stress plus a small increment.

The number of cycles of stress to failure (i.e., the fatigue life) of a component depends on the mean stress and the stress amplitude. Considering several identical components with the same stress history pattern yet with different mean stresses, the component with the largest mean stress will have the shortest fatigue life. If there is no dead load on the component, the mean stress will be nearly zero; and as such, the component will possess the longest possible fatigue life. In conducting the fatigue damage evaluation of a bridge component, the fatigue failure behavior (i.e., the stress range versus the number of cycles to failure curve, known as the S-N curve) needs to be determined using the Goodman diagram (Munse, 1990). In this study, the S-N curve for the bridge section with the cover-welded plate was obtained from Ang and Munse (1975). The S-N curve used herein does not include the mean stress effect. Structural analyses conducted on several of the example bridges under dead load alone showed that the dead load stresses at the fatigue-critical locations along bridge girders were very small. Hence, the S-N curve without the effect of the mean stress can be used with a reasonable level of accuracy.

Fatigue Damage Estimation

Fatigue damage is estimated from the Miner damage rule. Generally, a component is assumed failed when the cumulative damage computed from the Miner damage rule becomes equal to unity. In a probabilistic model the damage is treated as a random variable to account for various levels of uncertainties that may be inherent in the damage accumulation formulation. Sources of uncertainties can be attributed to: (i) stress data; (ii) Miner damage rule; and (iii) the S-N curve. Ang and Munse (1975) estimated a coefficient of variation equal to 0.54 to represent the uncertainty for steel girders with top and bottom cover plates. This value has been used in this paper. The estimation of fatigue damage and the remaining life can be achieved using either the procedure outlined by Moses et al (1987) or Ang and Munse (1975). Details of both methods can also be found in Byers et al (1997). The cumulative fatigue damage (D) is computed from the Miner rule by means of the following relationship.

$$D = n_1/N_1 + n_2/N_2 + \ldots + n_k/N_k \quad (1)$$

in which n_i (i=1,2, ..., k) is the number of cycles for the stress range S_i; N_i = total number of cycles to failure at stress range S_i; and k = total number of stress ranges. The parameter N_i can be obtained from the S-N curve. In this study, the following equation is used to represent this curve,

$$N = C/S^m \quad (2)$$

in which C and m are empirical constants. For the section with cover plates, $log_{10}C$ =8.6 and m=2.75 (Ang and Munse, 1975). Fatigue failure is assumed to occur when the expected value of damage, $E(D)$, approaches unity. The expected damage $E(D)$ is a statistical term which is equal to the statistical mean value of damage. Using Eqs. 1 and 2, it can be shown that:

$$E(D) = \int_0^{S_{max}} \frac{\bar{n}}{N(s)} f(s)ds \quad (3)$$

Equation 3 indicates that the stress range S is a continuous random variable which is described with $f(s)$ as the probability density function (the lower case s is an auxiliary variable that represents different values of the random variable S). Theoretical forms of $f(s)$ can be obtained using field stress range data (such as those depicted in Figs. 1-5) and tested for their validity using a statistical test such as the Chi Square test. In this study, a Beta probability density function is used for $f(s)$ which was validated for the stress range field data (see Figs. 1-5). Also in Eq. 3, $\bar{n}$ = average number of cycles (for all ranges) to failure and S_{max} = maximum stress range occurring in the component. Combining Eqs. 2 and 3 and with $E(D)$ = 1, one obtains:

$$\bar{n} = \frac{c}{\int_0^{S_{max}} S^m f(s)ds} \tag{4}$$

once $\bar{n}$ is computed, the reliability (i.e., the probability of no failure) of a component for a desired life of n cycles can be obtained from the Weibull distribution using Ang and Munse's (1975) equation:

$$L(n) = exp\left(-\left[\frac{n}{\bar{n}} \Gamma(1+\alpha) \right]^{1/\alpha} \right) \tag{5}$$

in which $L(n)$ = fatigue reliability, Γ= gamma function, $\alpha = \Omega^{1.08}$, Ω = uncertainty in fatigue life (0.54 for sections with cover plates), and $exp(.)=e^{(.)}$. In practice, one selects a desired reliability level, (for example, 97.7% for non-redundant bridges according to Moses, et al, 1987) and the fatigue life n is then computed. In this paper, upon computing n for fifteen example bridges, the data on the average daily traffic specific to the bridge roadway, distribution of traffic volume per traffic lane and expected traffic growth in future years are used to convert n into a unit of time (e.g., years). If t=fatigue life in years, then

$$n = (1+r)^t.\text{ADT}.(365).p \tag{6}$$

in which r = yearly rate of traffic volume increase, ADT=average daily traffic, and p= a factor corresponding to the distribution of the ADT per traffic lane. Equation 6 is then solved for t. The remaining fatigue life will be t-t_0 where t_0=the current age of the bridge in years.

Significance of Overloads

As described earlier, the compiled stress range data may not contain the stress ranges from overloaded trucks. Overloaded trucks use a bridge by permits that are issued by state transportation agencies. Mohammadi and Polepeddi (2000) discuss the significance of overloaded trucks in fatigue damage of bridge girders. While an infrequent application of an overload may not cause any significant fatigue damage to a bridge, many older bridges that are along highway feeder ramps may be dramatically affected. Most such bridges are designed for truck weights as low as 240 kN (54 kips), while permits for overloads are for trucks with gross weights in excess of 356 kN (80 kips). In a 25-year period, the extra amount of fatigue damage resulting from frequent overload applications may reach as high as 8% of the fatigue life of the bridge (Mohammadi and Polepeddi, 2000). This amount is based on an estimate of the population of overloaded trucks; and as such, may not necessarily represent a realistic account of fatigue damage due to overloads. Most state agencies keep records of overload permits. These records can be helpful in supplementing the data obtained in the field acquisition of the stress range values. However, a structural analysis of the bridge will be necessary to convert the overloaded truck gross weight data into stress range values. As reported in Mohammadi and Polepeddi (2000), a modification needs to be applied to the analytical stress range values to account for the difference between observed field stress data and those from analysis. This modification is reported to be in the 0.504-0.980 range.

Illustrative Examples

The practical application of the methods described in this paper is the estimation of remaining fatigue life of existing bridges and the condition assessment of bridges for the truck weight increase and/or traffic volume growth. One purpose of this study was to investigate the significance of truck weight increase and traffic volume growth to the service life of highway bridges. The field data compiled for the fifteen bridges provided the history of bridge stress ranges under current conditions. The investigation into the condition of these bridges was intended to specifically consider the following three factors and their effect on the remaining life of the bridges:

1. Truck weight increase by about 10% , from about 320 kN (72 kips) gross weight to 356 kN (80 kips),
2. Traffic volume growth on rural and urban highways, and
3. Overloaded truck population.

All three factors were expected to shorten the fatigue life of these bridges. The weight increase was considered to be a critical factor, especially for bridges that were designed for a lower gross truck weight. In addition to the weight increase, urban roadways experience an average of 5% traffic increase per year. The growth for rural roadways is about 5.6% per year (Mohammadi, Guralnick, Polepeddi, 1992). This growth will be compounded yearly and will accelerate the rate of fatigue damage growth. Considering the aforementioned two factors, the study followed the steps described below to arrive at the fatigue life of any given bridge:

- A target reliability level equal to 97.7% for non-redundant and 99.9% for redundant bridges was selected.
- Using the theoretical Beta distribution for the stress history of a bridge, an appropriate S-N relation, the mean life to failure for the critical components of the bridge was computed from Eq. 4. For non-redundant bridges, the occurrence of fatigue failure in a component is tantamount to severe bridge damage or failure. For a redundant bridge, failure in one or more components may cause some damage in the bridge varying from minor to severe. This damage may or may not necessarily cause the bridge to fail completely. However, for simplicity it was assumed that failure of a critical component in fatigue occurs at a time which is nearly identical to the fatigue life of the bridge. Using Eq. 5, the fatigue life n (i.e. the number of stress range cycles

to failure) was computed for two conditions; namely, without the truck weight increase and with the 10% weight increase.

- The rate of traffic volume increase and the average daily traffic for the roadway were used to convert the fatigue life into years. Appropriate factors to distribute the traffic volume to roadway lanes were used for this purpose.

- As an extension of this study, stress range data for overloaded trucks is used to estimate the percentage increase in fatigue damage for bridges that are frequently used by very heavy trucks (with gross weights in excess of 256 kN).

It is noted that the fatigue life so estimated is called the service life because it defines the time after which the serviceability of the bridge has been compromised because of unacceptable levels of fatigue damage to the critical components of the bridge.

To illustrate the applicability of this method as a non-destructive technique for bridge fatigue reliability assessment, a steel girder bridge is analyzed. This bridge is identified with the number 0420017. The bridge is a multi span (simply-supported) bridge carrying two lanes of traffic over the Macoupin Creek along US Route 67 in Illinois, USA. The bridge was originally built in 1941 with a major reconstruction completed in 1970. The average daily commercial traffic on the bridge is ADT=375. The bridge's overall length is 1310 feet (399 m); the longest span length=90 feet (27.5 m) and the bridge width=32.6 feet (9.9 m). The construction consists of a reinforced concrete slab supported by steel girders with welded cover plates. Bridge instrumentation consisted of two sets of strain gages located near the tip of the welded cover plates at the bottom flanges of the two girders. The stress range data obtained from field data is shown in Fig. 1 along with the theoretical values of the stress range. The traffic volume growth rate for rural bridges in Illinois is about 5.6%. Using Eqs. 4-6, the number of cycles to initiate fatigue cracking was computed as n=27,800,000 cycles using a reliability $L(n)$=97.7% in Eq. 5. Using r=5.6% and p=0.5, the estimated fatigue life=50 years. If no traffic volume growth is considered, the estimated fatigue life will be in excess of 75 years. Considering the reconstruction done in 1970, the current age in 2000 is taken to be 30 years. Hence, the remaining life in 2000 is estimated to be 20 years, if traffic volume growth is included in the analysis. In this illustrative example, it was assumed that the maximum truck weight remains unchanged. The study also considered the significance of 10% increase in the legal truck weight limit (increase from 320 to 356 kN) as discussed in the following section.

Discussion of Results

The results of a fatigue analysis of all bridges investigated in this study are summarized in Tables 1 and 2. Table 1 gives the results for the 72-kip (320 kN) weight limit; whereas Table 2 gives the results for the 80-kip (356 kN) weight limit.

Also given in Table 2 are the corresponding bridge lives with and without future traffic growth. It is emphasized that the results shown in these tables are given only for the purpose of assessing the effects of truck weight increase and traffic growth. Therefore, the estimated life (and not the remaining life) was used. Figure 6 depicts a summary of bridge life results considering the effect of traffic volume growth only. The 356 kN (72-kip) gross truck weight limit is used for this analysis. Bridges are then divided into 3 groups, those with lives shorter than 25 years, those with lives between 25 and 75 years and those with lives longer than 75 years. The 75-year life is considered in design to be the maximum useful bridge life. Hence, any fatigue life longer than 75 years will be limited by other factors including normal wear, deterioration, etc. As seen in Fig. 6, with 5.6% traffic growth, more bridges are tall into the group with shorter fatigue life and especially into the shorter-than-25-year group, than is the case without traffic growth. Figure 7 shows similar results with the 320-kN gross truck weight.

As reported in Mohammadi, et al (1998), considering a gross truck weight increase from 320 to 356 kNs, the results shown in Fig. 8 are obtained. Traffic growth is considered for both weight limits in obtaining the results of Fig. 8. As seen in this figure, with the 320-kN limit, 53% of bridges had a fatigue life longer than 75 years. With the 356-kN limit, the percentage of bridges with a life shorter than 25 years has increased by 6% (from 14% to 20% as shown in Fig. 8). Thus it can be concluded that the weight increase alone does not dramatically affect these bridges.

Figure 9 compares fatigue lives with 320-kN gross truck weight but without traffic growth with corresponding to those with 356-kN gross truck weight but <u>with</u> traffic growth. An examination of the results shown in this figure reveals that with the 320-kN weight and no traffic growth, nearly 80% of the bridges have a fatigue life longer than 75 years. However, with the 356-kN gross truck weight and <u>with</u> predicted traffic growth, only 40% of the bridges will have a life of 75 years or longer. The results indicate that 60% of these bridges will have a life shorter than 50 years and, in particular 20% will have a life shorter than 25 years (see Fig. 9). From the results of Fig. 6-9, it may be concluded that an annual traffic growth of 5.6% is a more influential parameter in reducing bridge life than is an increase in gross truck weight (Mohammadi, et al, 1998).

The significance of overloaded trucks on fatigue life can also be investigated using Eqs.1-5. However, the stress range data specific to overloaded trucks must be acquired. One method to do this is by obtaining the statistics of overloaded trucks from state records and conducting structural analyses. This method was used by Mohammadi and Polepeddi (2000). They report that the overloads may increase the fatigue damage by about 8% over a 25-year period. The estimated fatigue damage obtained may be used as a measure to curb the total number of overload permits in an effort to cut on the overall cost of maintaining bridges.

Bridge Life Cycle Cost Analysis

The results of fatigue reliability estimation can be used in a comprehensive bridge life cycle cost analysis. Using an accepted reliability level for a given bridge, the estimated remaining fatigue life can be used as a base-line for the bridge life cycle cost analysis. Considering t as the remaining life, one can assume this as the time when the bridge is scheduled for replacement. Alternatives for bridge repair and replacement may include: (1) continue using the bridge until such time when bridge must be replaced; and (2) repair at an earlier time to extend the bridge useful life. If within time t, a change in the loading regime or traffic volume on the bridge occurs, the estimated life may be shortened. Combining this change with the two alternatives, several possible scenarios may be considered for the bridge. These may range from the "do nothing" approach to the "repair now" strategy. An engineering economics analysis will then be necessary to identify the most cost-effective approach. In this paper, the economics analysis associated with a change in the loading regime and traffic volume change on highway bridge is presented.

When a bridge permissible growth weight is increased by 10% and the traffic volume grows by about 5% annually, the fatigue life is shortened. The results from this study show that without the weight increase and traffic growth, the average remaining life of the bridges investigated is about 25 years. With the weight increase and traffic growth, the remaining life will be only 15 years. An ordinary bridge of moderate span length with two lanes of traffic and steel-girder construction with reinforced concrete deck slab costs about \$1,490,000 to replace. Considering this as the current replacement cost, the future cost of the bridge replacement after implementation of weight increase and traffic growth and using an inflation rate of 4% is

$$(FV)_{\text{future}}=1{,}490{,}000(1+0.04)^{15}=2{,}683{,}405$$

With 5% interest rate, we can compute the annualized cost for the future value which is

$$(P)_{N=15}=\frac{2{,}683{,}405\times 0.05}{1.05^{15}-1}=144{,}761$$

If no weight increase and traffic growth is considered, the remaining life is 25 years. The sequence of calculations in this case is

$$(FV)_{\text{future}}=1{,}490{,}000(1+0.04)^{25}=3{,}972{,}096$$

$$(P)_{N=25}=\frac{3{,}972{,}096\times 0.05}{1.05^{25}-1}=83{,}225$$

Thus the annual cost for the weight increase and traffic growth is 144,761-83,225=$61,536 per bridge. This is the cost associated with early bridge replacement due to the weight increase and traffic growth.

Summary and Conclusions

This paper presents the application of stress range data as a non-destructive method for estimating the fatigue reliability of highway bridges. The field data has been used to drive the probability density functions for the stress range data at fatigue critical locations in the steel girders of the example bridges. The functions thus obtained may be used, along with critical component S-N curves and a probabilistic method, to estimate fatigue life. Fatigue lives in years can be estimated from information on the average daily commercial traffic (ADT) on the bridge, the rate of increase in the traffic volume and the distribution of ADT to the various traffic lanes. The method outlined herein was used for several bridges in Illinois to analyze the significance of increases in legal truck weight and traffic volume upon bridge fatigue lives. As a result of this study, it is concluded that:

- Field stress range data can be used effectively and easily to estimate fatigue life of bridges.
- The duration of the data acquisition period may be limited to only a few days to capture the most significant variations in the loads experienced by a bridge.
- The method described herein for fatigue life estimation can be useful in evaluating effects upon bridge life when changes in legal truck weight and/or other factors such as traffic increases are expected.
- Fatigue reliability results and the estimated remaining lives of bridges can be utilized in a bridge life cycle cost analysis in conjunction with a comprehensive bridge management system.
- The fatigue reliability assessment method can easily be extended to incorporate the significance of overloaded trucks using a given bridge.

References

Ang, A. H.-S. and Tang, W.H. (1975). *Probability concepts in engineering planning and design, volume I-basic principles*, John Wiley and Sons, New York, New York.

Ang, A. H.-S. and Munse, W.H. (1975). "Practical reliability basis for structural fatigue." *Preprint No.2494, Proc., Nat. Struct. Engrg. Conf.,* ASCE, New York, N.Y.

Byers, W.G., Marley, M.J., Mohammadi, J., Nielsen, R.J., and Sarkani, S. (1997). "Fatigue reliability reassessment applications: state-of-the-art paper." *J. Struct. Engrg.,* ASCE, 123(3), 277-285.

Hahin, C., South, J.M., Mohammadi, J., and Polepeddi, R. (1993). "Accurate and rapid determination of fatigue damage in steel bridges." *J. of Struct. Engrg.*, ASCE, 119(1), 150-168.

Laman, J.A., and Nowak, A. (1996). "Fatigue-load model for girder bridges." *J. of Struct. Engrg.*, ASCE, 122(7), 726-233.

Mohammadi, J., Guralnick, S.A., and Polepeddi, R. (1992). "Effect of increased truck weight upon Illinois highway bridges." *Rep. No. FHWA/IL/RC-013*, Dept. of Civil Engineering., Illinois Institute of Technology, Chicago, IL.

Mohammadi, J., Guralnick, S.A., and Polepeddi, R. (1998). "Bridge fatigue life estimation from field data," *Practice Periodical on Structural Design and Construction*, ASCE, Vol. 3, No. 3, p. 128-133.

Mohammadi, J., and Polepeddi, R. (2000), "Bridge rating with consideration for fatigue damage from overloads," *J. of Bridge Engineering*, ASCE, 4(3).

Moses, F., Schilling, C.G., and Raju, K.S. (1987). "Fatigue evaluation procedures for steel bridges." *Nat. Cooperative Hwy. Res. Program, Rep. No. 299,* Transportation Research Board, National Research Council, Washington, D.C.

Munse, W.H. (1990). "Fatigue, brittle fracture and lamellar tearing." *Structural engineering handbook, 3rd Ed.*, E. H. Gaylord Jr. and C. N. Gaylord, eds., McGraw-Hill Book Co., 4.1-4.24

TABLE 1

Estimated Fatigue Lives (years) for 320 kN (72Kip) Truck Weight Limit

Bridge #	With 5.6% Traffic Growth	Without Traffic Growth
040031	73	≥ 75
0420017	51	≥ 75
0650005	50	≥ 75
0720014	45	≥ 75
0570050	≥ 75	≥ 75
0600126	≥ 75	≥ 75
0570067	≥ 75	≥ 75
0570068	≥ 75	≥ 75
0160335	14	21
1010019	≥ 75	≥ 75
1010056	≥ 75	≥ 75
0160158	18	35
0990055	11	17
0161022	≥ 75	≥ 75

TABLE 2

Estimated Fatigue Lives of Example Bridges (years) for 356 kN (80Kip) Truck Weight Limit

Bridge #	With 5.6% Traffic Growth	Without Traffic Growth
0540031	73	≥ 75
0420017	50	≥ 75
0650005	47	67
0720014	44	51
0570050	64	≥ 75
0600126	48	72
0570067	≥ 75	≥ 75
0570068	≥ 75	≥ 75
0160335	13	21
1010019	≥ 75	≥ 75
1010055	≥ 75	≥ 75
1010056	≥ 75	≥ 75
0160158	18	34
0990055	11	17
0161022	≥ 75	≥ 75

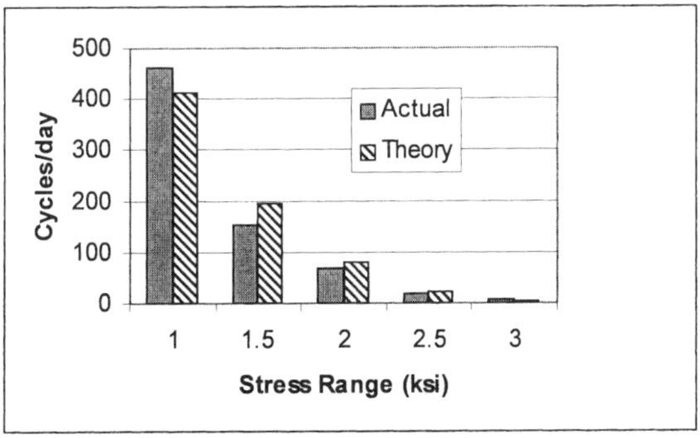

Fig 1. Stress Range versus Frequency for Bridge 0420017

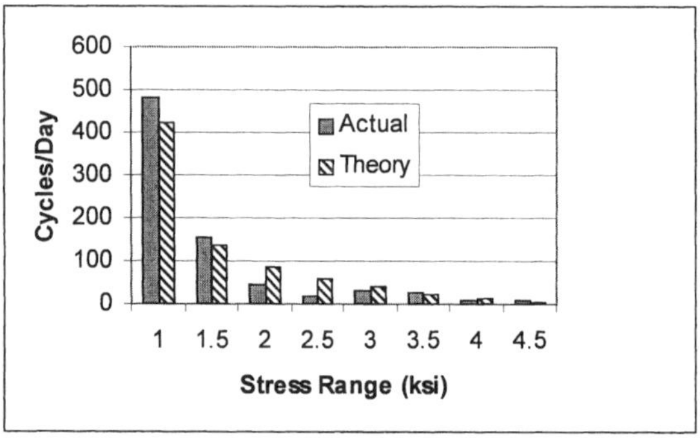

Fig. 2. Stress Range versus Frequency for Bridge 0650005

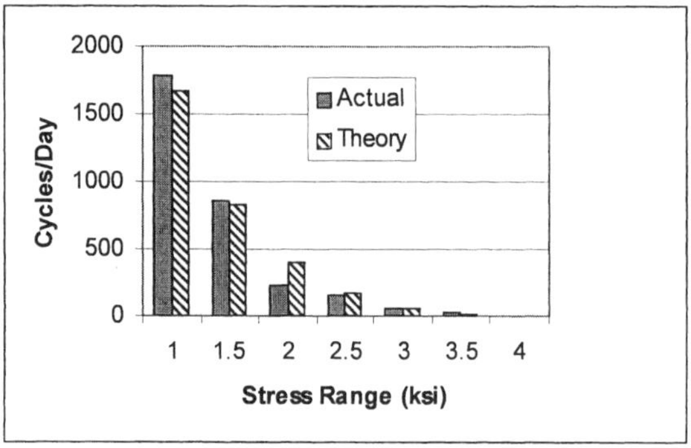

Fig. 3. Stress Range versus Frequency for Bridge 1010019

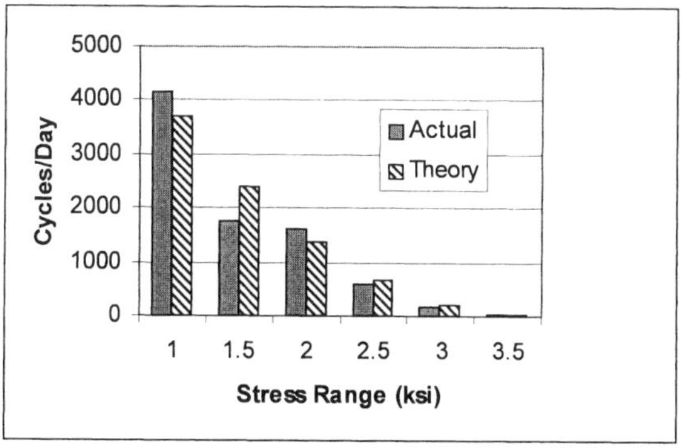

Fig. 4. Stress Range versus Frequency for Bridge 0990055

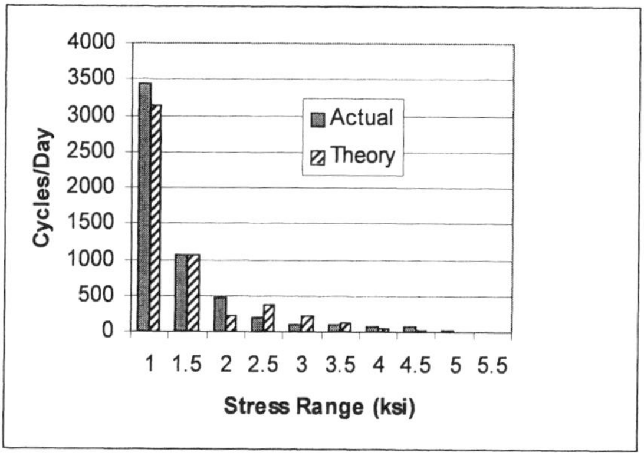

Fig. 5. Stress Range versus Frequency for Bridge 0160335

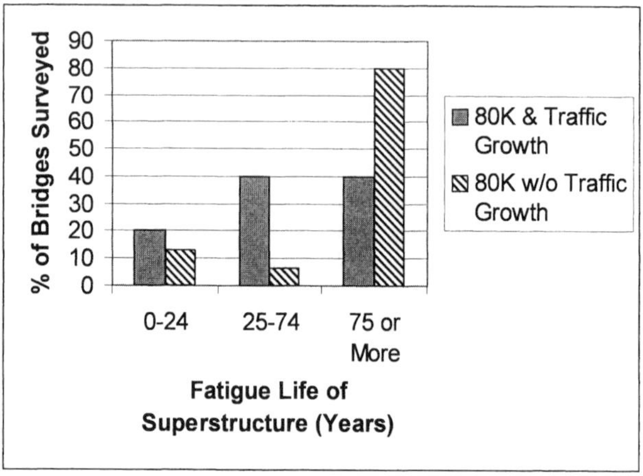

Fig. 6. Effect of Traffic Growth on Fatigue Life (80 K Growth Weight)

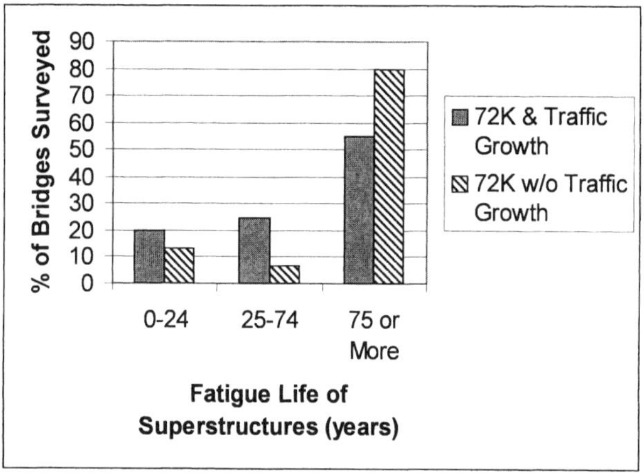

Fig. 7. Effect of Traffic Growth on Fatigue Life (72 K Gross Weight)

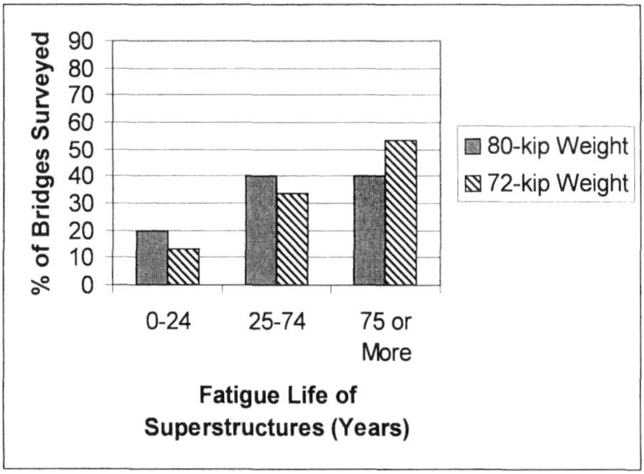

Fig. 8. Effect of Gross Truck Weight Increase on Fatigue Life

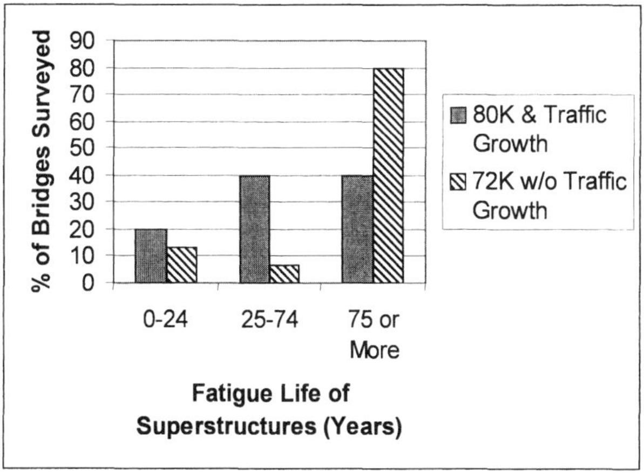

Fig. 9. Combined Effect of Gross Truck Weight Increase and Traffic Growth on Fatigue Life

Nondestructive Evaluation of Residual Stresses in Welded Joints with Application to Stochastic Fatigue Damage Accumulation

Shahram Sarkani[1], George Michaelov[2], and Loren D. Lutes[3]

Abstract

Some of the most popular techniques for nondestructive evaluation of residual stresses in welded joints are described. Particular attention is paid to the X-ray diffraction and the neutron diffraction method. The neutron diffraction method is the only nondestructive technique that is able to provide a complete, through-thickness distribution of the residual stresses in welded joints. The information on residual stresses is particularly important for fatigue damage calculation under stochastic loadings. It is demonstrated that the influence of residual stresses on stochastic fatigue damage accumulation may be incorporated by a simple approach based on elastic-perfectly-plastic material model and the Gerber correction factor. The model assumes that the remaining residual stress at the critical location depends on the largest nominal stress ever seen by a welded joint. The model predicts that the residual stresses during stochastic loading randomly decay to zero. The effect of material yielding is additionally investigated by considering an elastic-plastic material model with linear kinematic hardening. The residual stresses in this case are computed through Monte-Carlo simulations. It is demonstrated that the effect of material hardening is to reduce the rate of residual stress decay and thus to accelerate the rate of fatigue damage accumulation.

Introduction

Residual stresses are those internally self-balanced stresses that remain in a structure or component after any external loading has been removed. They usually

[1] School of Engineering and Applied Science, The George Washington University, Washington, DC 20052
[2] Pragma Systems Corporation, Reston, VA 20190
[3] Department of Civil Engrg, Texas A&M University, College Station, TX 77098

occur as a result of non-uniform plastic deformations taking place during manufacturing processes such as rolling, forging, casting, and welding. Residual stresses may change over time as a result of the service loadings experienced by the structure. Tensile residual stresses are usually considered undesirable because they promote formation and propagation of cracks. For this reason, the presence of compressive residual stresses is usually considered desirable: some manufacturing techniques like shot peening are often employed that deliberately introduce large compressive residual stresses on the surface to inhibit surface crack formation.

Accurate quantification of residual stresses is important for various engineering analyses. Unfortunately, the processes that lead to residual stress formation may be extremely complex and difficult to model analytically. This is particularly true for residual stresses due to welding. Although considerable progress into the finite element modeling of the welding processes has been achieved, such analyses are still extremely computationally involved and generally suffer from lack of reliable data on material properties at high temperature (Goldak 1989). That is why engineering practitioners have paid considerable attention to experimental methods for residual stress determination and particularly to those that allow such determination to be carried out in a nondestructive manner.

Fatigue is one of the material failure modes that may be significantly influenced by residual stress fields. The residual stresses and the external loading in a welded joint interact in a complex manner. From a fatigue analysis perspective, the primary result of this interaction is the occurrence of mean stresses in the local stress cycles residing in the vicinity of the welded joint regardless of whether or not the nominal loading on the joint is mean zero. Mean stresses in general do not represent significant difficulties for fatigue calculations—in the presence of zero-mean constant-amplitude data their effect may be accounted for by empirical corrections of the stress range like those developed by Goodman and Gerber more than a century ago (Fuchs and Stephens 1980). Moreover, if the mean stresses under stochastic loadings are time-invariant and if the same constant mean values can be maintained during constant-amplitude testing, one need not worry at all about mean stress. Its effect is implicitly included in the S-N curve, so it is automatically reflected in the predicted stochastic fatigue damage. However, mean stresses resulting from cyclic loading and initial residual stress in welded joint made of elastic-plastic material are neither constant over time nor do they vary the same way under stochastic as they do under constant-amplitude loadings.

This study focuses on fatigue and residual stresses of welded structures with two main objectives. The first is to summarize some of the nondestructive test methods that may be used for residual stress determination. The study particularly focuses on two diffraction techniques: X-ray and neutron. The second objective is to provide a fatigue analyst with practical guidelines regarding how the information gathered by the nondestructive techniques may be used to quantify the residual stress effects, particularly in stochastic fatigue damage accumulation.

Nondestructive Evaluation of Weld Residual Stresses

Although the term stress measurement has come to common usage, stress is an extrinsic property that is not directly measurable. All methods of stress determination require measurement of some intrinsic property, such as strain, and the calculation of the associated stress. The nondestructive methods for residual stress measurement can generally be classified into the following categories: mechanical methods, nonlinear elastic methods (ultrasonic and magnetic), and diffraction methods (X-ray and neutron).

The most popular mechanical method is the hole drilling technique. It is based on the principle that if a portion of a material is relieved of its stress, it will expand or contract depending upon the stress state. The standard technique consists of mounting strain gages on the area to be investigated. After the hole is drilled, the strain resulting from the stress release is measured and converted to stress by the use of general Hooke's law. The equipment required to perform the measurements is relatively inexpensive and is readily available on the market along with specially tailored strain gages. Instead of strain gages, holographic interferometers can be used, which allows for smaller holes to be drilled but qualitative analysis may be more difficult. A more recent modification of the hole drilling method is the ring drilling method. A ring of material is cut around the strain gage, thereby measuring the residual stress at the point of interest directly. This ring method generally increases the amount of surface residual stress relieved thus providing for better resolution. Another mechanical method is the indentation method. This technique uses local deformations of the surface (similar to those used in hardness measurements) and measures the displacements of the surface after releasing the indention, which are correlated to the surface residual stresses. The surface displacement is usually measured by holographic, Moire, or speckle interferometry. The major advantage of the mechanical methods is in their low cost and in obtaining results quickly even in field conditions. The major disadvantage is that the mechanical methods are not strictly nondestructive techniques because they leave certain damage on the measured structure. Additionally, they are usually unable to account for high stress gradients since the gage (sampling) volumes are relatively large.

The ultrasonic methods are based on the acoustoelastic effect, which correlates the velocity of an acoustic wave to the stress in a solid. Magnetic methods use similar correlation between magnetic permeability and stress. Both the ultrasonic and magnetic methods have received limited practical application due to significant errors and limited sensitivity of the measurements. Both require stress-free reference samples, which are identical to the sample investigated.

The diffraction methods have been established as one of the most reliable and accurate residual stress measurement techniques. A diffraction method is probably the most direct approach to measure strain inside a material. It is based on the fact that the spacing of the atomic planes in polycrystalline materials, such as steels, is very well defined at any given temperature and is altered by stress. The change in spacing is determined by measurement of a diffraction angle and is subsequently related to the existing stress. There are two major types of diffraction methods based on the irradiation source: X-ray and neutron. The X-ray diffraction technique is the older of the two methods. Commercial instruments have been developed for both laboratory and field measurement. Reports on application of neutron diffraction to residual stress measurement first appeared in the mid 80's. The technique requires an intense source of neutrons. Currently such sources exist at nuclear fission reactors and particle accelerators. (Neutron diffraction capabilities to measure residual stress exist at the Oak Ridge National Laboratory in the USA and at the Chalk River Laboratories in Canada.) Thus, a major disadvantage of the neutron technique is its non-portability – objects to be investigated must be brought to the neutron source. The biggest advantage of the neutron technique over the X-ray is the high penetration potential of neutrons. Thus, while the X-ray method can provide stress distribution near the material surface, the neutron method can provide a whole three-dimensional mapping of the residual stress state in the component interior. Weldments of ferritic steel, stainless steel, and of nickel zirconium and titanium alloys have all been studied by neutron diffraction (Holden et al. 1988; Holden 1995, Holden et al. 1995, Klassen, et al. 1991; Root et al. 1993; Mahin et al. 1991; Jo et al. 1993).

Both the X-ray and the neutron diffraction method make use of Bragg's law

$$\lambda = 2d\sin(\theta) \tag{1}$$

in which λ is the wavelength of the source, d is the distance between the atomic planes of the crystal lattice, and 2θ is the diffraction angle, i.e., the angle between the incident and diffracted (scattered) beams. The region of intersection of the incident and scattered beams is the gage volume, which gives the desired spatial resolution of the measured strains. Strain is measured in the direction that bisects the incident and scattered beams. This strain is computed as

$$\varepsilon = \frac{d - d_0}{d_0} \tag{2}$$

in which d_0 is the stress-free lattice spacing.

The X-ray and neutron diffraction techniques differ in the methods used to calculate residual stresses. Since the X-ray diffraction method deals with a surface layer, a condition of plane stress is assumed to exist. The neutron diffraction technique is not bounded by such limitations and usually a true tri-axial strain state is assumed. By appropriately rotating the incident-scattered beams, the bisector is oriented at different directions, which allows the strains in three perpendicular directions to be measured. With three measurements of strain, the residual stresses are calculated through the generalized Hooke's law

$$\sigma_i = \frac{E}{1+\nu}\left[\varepsilon_i + \frac{\nu}{1-2\nu}\left(\varepsilon_i + \varepsilon_j + \varepsilon_k\right)\right] \tag{3}$$

in which E is the Young's modulus, ν is the Poisson ratio, and the subscripts indicate the components of stress and strain.

Note that the linear elastic relations (3) hold even when plastic strains are present at the residual state. The change in lattice spacing is related only to the elastic (recoverable) deformations. In addition, plastic flow theory usually postulates the existence of a stress-free state to which the material can be transformed by pure elastic deformations.

Effect of Residual Stresses on Stochastic Fatigue Damage Accumulation

Stochastic fatigue damage accumulation refers to the situation when fatigue damage takes place under variable-amplitude loading modeled as a stochastic (random) process. Calculation of fatigue damage under such conditions is usually performed on the premise that a single cycle in the variable-amplitude loading causes the same amount of damage as a cycle with the same magnitude in a constant-amplitude loading. This postulate, which is commonly known as the linear damage accumulation rule or the Palmgren-Miner hypothesis (Palmgren, 1924; Miner 1945), provides a sufficiently flexible framework for fatigue life estimation under a number of different conditions. The two basic elements of the calculation strategy are the technique for identifying the fatigue-inducing cycles and S-N data giving the constant-amplitude fatigue lives of representative specimens. Among the cycle-identification techniques the rainflow method is probably the most accepted general algorithm but it requires simulation of sample histories of the stochastic loading. A simple technique directly based on the stochastic process theory is the Rayleigh approximation but it is only suitable for narrowband loadings. A number of "correction factor" approaches to extend the Rayleigh approximation to broadband loadings have been proposed in the 80s. More recent techniques applicable to broadband situations are those of Bishop and Sherratt(1990), which uses an approximation of the probability distribution of rainflow cycles, and the single-moment method of Lutes and Larsen (1990).

A relatively simple model to take into account the mean stress effects due to residual stresses in stochastic fatigue damage accumulation has been developed by Lutes and Sarkani (1996). The model assumes that the fatigue failure of a welded joint is controlled by the local stress history at a "critical location" (usually the weld toe with stress concentration), at which point the initial residual stress equals the material's yield stress. By using an elastic-perfectly-plastic material assumption, a simple analytical relationship between the local stress history and the nominal loading on the joint is derived. The result is that the mean stress at the critical location is a function of the largest load ever applied to the joint. As a result, under constant amplitude loading the mean stress developed after the first cycle stays constant while under stochastic loading it gradually decays in a random manner throughout the fatigue life of the joint. Recent studies in Sweden (Blom 1995) have experimentally confirmed the decay of residual stress in welded joints under stochastic loading by performing repetitive X-ray diffraction measurements through the fatigue tests.

The simple analytical relationship between nominal loading and local stress assumed by Lutes and Sarkani may hold only if the material is elastic-perfectly plastic and has a one-dimensional stress state, i.e., if the maximum stress at the critical location cannot exceed the yield stress. That may not be the case in real joints where material hardening or three-dimensional confinement may result in stresses higher than yield. In such a case, the stresses at the critical location should be determined from the response of the joint as a whole. Note, however, that analyzing the joint based on a more complex plasticity model is useless if only the residual stress on the surface of the weld toe is known, which is the case for most NDT methods, like X-ray diffraction.

As already discussed, neutron diffraction is capable of rendering a more complete picture of the through thickness and three-dimensional nature of the residual stress distribution. It is the objective of this study to demonstrate how this information may be utilized in fatigue damage calculations. In addition, the effect of material strain hardening on the decay rates of the residual stresses is investigated. The analysis presented is an extension of the original Lutes and Sarkani model (1996). However, it is computationally much more involved than the original model because computing the statistics of the local stress histories from the statistics of the nominal loading in the presence of strain hardening requires the utilization of Monte-Carlo simulations. For this reason, it is also expedient to compare the results rendered by the complex elastic-plastic, strain-hardening model with those of the simple original model.

Welded Joint Considered

As in the original Lutes and Sarkani (1996) investigation, the specific example considered in this study is a tee specimen formed by joining two 25.4-mm thick plates with full-penetration fillet welds. The T-weld is shown in Fig. 1. The specimen is loaded through the vertical plate while the two ends of the horizontal plate are held simply supported. The yield and the ultimate stress of the material are taken as σ_y=345 MPa and σ_{ult}=1.55σ_y, respectively. The Young's modulus was assumed to be E=204,000 MPa. Two different material models are considered: elastic-perfectly-plastic and elastic-plastic with linear kinematic hardening. For the model with hardening, the slope of the stress-strain curve above yield, the plastic modulus, is assumed to be E_p=E/20. The material models are shown in Fig. 2. The joint is modeled as an elastic-plastic beam, the deformations of which obey the Bernoulli-Navier beam theory. In other words, it is assumed that the plane sections remain plane and perpendicular to the X-axis during rotation and that the total bending strains (the sum of elastic and plastic parts) follow a linear distribution proportional to that rotation. The mode of loading on the joint puts the critical location for fatigue cracking at the weld toe of the horizontal plate. Accordingly, the nominal stress for fatigue damage calculations is taken as the elastic stress (bending moment divided by section modulus) acting on the weld toe.

Note that residual stress fields on welded joints typically exhibit high stress-triaxiality. (See, for example, the neutron diffraction results of Holden et al. (1988) on a T-weld with geometry and thickness similar to those assumed in this study.) For simplicity, the following analysis ignores the stress-triaxiality effects and considers only residual stresses parallel to the X-axis. It is assumed that neutron diffraction scanning of the cross-section under the weld toe has rendered the distribution shown in Fig. 3. This distribution represents a cubic polynomial with a maximum at the weld toe equal to σ_y. Residual stresses with similar distribution have been obtained by Holden et al. (1988) on a T-weld with a similar geometry and thickness.

Stochastic Fatigue Modeling

Prior studies (Sarkani and Lutes 1988; Kihl and Sarkani 1996) have shown that the Gerber formula can be effectively used to correct for the effect of mean stress in constant-amplitude tests. The Gerber technique provides for an equivalent stress range which for a given loading cycle with stress range (double amplitude) S and mean stress σ_{mean} is expressed as

$$S_{\text{eff}} = S\left[1-\left(\frac{\sigma_{mean}}{\sigma_{ult}}\right)^2\right]^{-1} \tag{4}$$

The constant-amplitude S-N curve is then taken in the usual power-law form, but based on S_{eff} instead of the actual stress range S; that is, the cycles to failure are computed from

$$N = KS_{\text{eff}}^{-m} \tag{5}$$

where K and m are known as the fatigue life coefficient and the fatigue strength exponent, respectively.

According to the Palmgren-Miner rule, the fatigue damage ΔD_j during cycle j having equivalent stress range S_{eff} is taken as

$$\Delta D_j = K^{-1}S_{\text{eff}}^{m} \tag{6}$$

Substituting (4) into (6) gives

$$\Delta D_j = K^{-1}S^{m}\left[1-\left(\frac{\sigma_{mean}}{\sigma_{ult}}\right)^2\right]^{-m} \tag{7}$$

The above equation shows that to obtain the damage per cycle with non-zero mean stress one needs to calculate the value predicted without mean stress, then multiply the result by a correction factor of

$$CF = \left[1-\left(\frac{\sigma_{mean}}{\sigma_{ult}}\right)^2\right]^{-m} \tag{8}$$

Figure 4 shows a plot of the correction factor as function of the mean stress for three possible values of m (the range from 3 to 5 is quite common for the fatigue exponent of welded joints). The plots is for the particular case $\sigma_{ult}=1.55\sigma_y$ assumed in this study. It can be seen that the correction factor can increase significantly for larger mean stress values.

In this study it is assumed that the nominal stress on the joint represents a stationary, narrowband, zero-mean, Gaussian process $\sigma(t)$ with root-mean-square (RMS) of σ_{rms}. Accordingly, the probability distribution of the stress amplitude σ_{ampl} is Rayleigh and can be expressed as

$$p_{\sigma_{ampl}}(x) = \frac{x}{\sigma_{rms}^2}\exp\left(-\frac{x^2}{2\sigma_{rms}^2}\right) \tag{9}$$

For a narrowband process the stress range is twice the amplitude, i.e. $S=2\sigma_{ampl}$.

Let $D(t)$ denote the accumulated damage after $n(t)$ cycles with the above distribution. The expected value of accumulated damage $E[D(t)]$ may be found from Eqs. (4), (5), and (6) as

$$\begin{aligned} E[D(t)] &= \sum_{j=1}^{n(t)} E(\Delta D_j) = K^{-1}E(S^m)\sum_{j=1}^{n(t)} E(CF_j) = \\ &K^{-1}2^{3m/2}\sigma_{rms}^m\Gamma\left(1+\frac{m}{2}\right)\sum_{j=1}^{n(t)} E(CF_j) \end{aligned} \tag{10}$$

The predicted fatigue life for a given value of σ_{rms} is obtained by finding the value of $n(t)$ such that $E[D(t)]=1$. That is, the predicted life is the value n such that

$$\sum_{j=1}^{N} E(CF_j) = K\left[2^{3m/2}\sigma_{rms}^m\Gamma\left(1+\frac{m}{2}\right)\right]^{-1} \tag{11}$$

If there were no residual stresses then CF_j would be unity in every cycle and the left hand side of Eq.(11) would simply be n. For CF_j different from unity, Eq.(11) represents a nonlinear equation for N, which needs to be solved by iterations. The values of $E(CF_j)$ for each cycle will depend on the material model and the initial residual stress field. Evaluation of $E(CF_j)$ may be quite computationally involved in general. For elastic-perfectly-plastic material, however, a relatively simple approximation can be devised. This approximation and corresponding stochastic fatigue life prediction are discussed in the following section

Decay of Residual Stress for Elastic-Perfectly-Plastic Material

If the material of the joint is elastic perfectly plastic, the local stress response at the weld toe can be obtained independently from the entire cross-section. That is, the response obtained by considering the weld toe as an elastic-plastic material point subjected to nominal stress $\sigma(t)$ is the same as the response obtained when the entire cross-section is subjected to a nominal bending moment of magnitude $\sigma(t)W$, where W is the elastic section modulus. This makes the elastic-perfectly-plastic assumption very suitable for situations when only the residual stress at the surface is known (e.g., X-ray diffraction measurements).

Consider now the effect on the stress at the weld toe of applying then removing a load causing the nominal stress to increase from zero to σ_{max} then return to zero. First consider the $\sigma_{max} < \sigma_y$ so that the overall behavior is linear. Considering that the magnitude of the initial residual stress is σ_y and that the local stress cannot exceed σ_y, the initial loading causes yielding with no increase in stress, and the unloading causes the local stress to drop by an amount σ_{max} to the level σ_y - σ_{max}. Subsequent cycling of the nominal stress with a mean value of zero and magnitude equal to σ_{max} causes the local stress to vary about the level σ_y - σ_{max}. Thus, a zero mean nominal loading gives $\sigma_{mean}=\sigma_y-\sigma_{max}$ as long as the amplitudes of the subsequent loading cycles do not exceed σ_{max}. If $\sigma_{max} > \sigma_y$ more general yielding in the specimen occurs so that the unloading strain is no longer proportional to σ_{max}. It can be assumed that the unloading strain is proportional to σ_y so that the effect of loading and unloading is to reduce the local residual stress to zero when $\sigma_{max} > \sigma_y$. In a summary, we have $\sigma_{mean}=\sigma_y-\sigma_{max}$ for $\sigma_{max} < \sigma_y$ and $\sigma_{mean}=0$ for $\sigma_{max} > \sigma_y$.

For constant-amplitude loading testing with mean zero and amplitude σ_{max} the redistribution of residual stresses described above occurs during the first cycle. That is, the mean (and for that matter the residual) stress at the critical location would be $\sigma_{mean}=\sigma_y-\sigma_{max}$ for $\sigma_{max} < \sigma_y$ or $\sigma_{mean}=0$ for $\sigma_{max} > \sigma_y$ for all cycles. For stochastic loadings there may be many events of mean (residual) stress change. These events take place any time a local yielding at the weld toe occurs. Such local yielding occurs anytime that the amplitude of the nominal stress becomes larger than it has ever been before. As a result, the mean (residual) stress would gradually decay until reaching zero at a cycle with amplitude $\sigma_{max} \geq \sigma_y$ and would stay at the zero level through failure. The mean stress at any time depends on the largest value of nominal stress σ_{max} at any time in the past (Sarkani and Lutes 1988). For a stochastic loading, this σ_{max} is a random variable depending on the past history of the loading.

Based on the above residual stress model, the correction factor for the jth cycle of stochastic stress can be expressed as

$$CF_j = \left[1 - \left(\frac{\sigma_y - \sigma_{max}}{\sigma_{ult}}\right)^2\right]^{-m} \quad \text{for} \quad \sigma_{max} < \sigma_y$$
$$CF_j = 1 \quad \text{for} \quad \sigma_{max} \geq \sigma_y \tag{12}$$

where σ_{max} is the maximum nominal stress amplitude for all cycles $n(t) \leq j$. Thus, to evaluate the expected value of the correction factor, required by Eq.(11), one needs to know the probability distribution of the random σ_{max} – the extreme value of the stochastic nominal stress process.

Although the exact probability distribution of the extreme value of a stochastic process has not been derived, a number of approximations exist. The simplest of these is the so-called "Poisson approximation," which is based on the assumption that the upcrossings of the process are statistically independent events and thus occur according to the Poisson distribution. This gives the cumulative distribution of σ_{max} in the interval $[0, t]$ as

$$P[\sigma_{\max} \le x] = P[\sigma(0) \le x]\exp\left[-n(t)\exp\left(-\frac{x^2}{2\sigma_{rms}^2}\right)\right] = \left[1-\exp\left(-\frac{x^2}{2\sigma_{rms}^2}\right)\right]\exp\left[-n(t)\exp\left(-\frac{x^2}{2\sigma_{rms}^2}\right)\right] \tag{13}$$

Strictly speaking in the above equation the Gaussian distribution should have been used for the initial probability of upcrossing level x, $P[\sigma(0)\le x]$. Instead, Eq.(13) incorporates the Rayleigh distribution for this initial event, which seems to improve the approximation for $P[\sigma_{max}\le x]$. This improvement can be by the fact that if the amplitude of the first cycle is above level x but $\sigma(0)$ happens to be less than x, there is very high probability that $\sigma(t)$ crosses the level x within the first cycle. Thus using the Rayleigh distribution for the initial condition term in (13) provides a correction for this dependence between the initial value of the process and the upcrossing times, which is not accounted for by the independent upcrossings approximation. In practice, the problem of what $P[\sigma(0)\le x]$ to use does not seem to be very important for fatigue damage accumulation. It may be shown that the probability distribution $P[\sigma_{max}\le x]$ becomes almost independent of the initial conditions after a few tens of cycles. Very little fatigue damage may be expected to occur during these few cycles and one may simply assume $P[\sigma(0)\le x] = 1$.

An analytical approximation for the expected value $E[\sigma_{max}]$ based on Eq.(13) was derived by Davenport (1964) as

$$E(\sigma_{\max}) \approx \left(\sqrt{2\ln[n(t)]} + \frac{0.577}{\sqrt{2\ln[n(t)]}}\right)\sigma_{rms} \tag{14}$$

This approximation asymptotically approaches the exact $E(\sigma_{max})$ for a large number of cycles. Its largest discrepancy is for $n(t)=1$ for which Eq.(14) yields infinity.

The Poisson crossing approximation, on which Eqs.(13) and (14) are based, is known to yield very good results when the stochastic stress process $\sigma(t)$ is broadband or the upcrossing level x is high. Only under such conditions will the uprossings be independent events. For the case of a narrowband stress process $\sigma(t)$, a better approximation of $P[\sigma_{max}\le x]$ accepted widely is the expression derived by Vanmarcke (1975). Based on the Vanmarcke approximation, Der Kiureghian (1980) derived an approximation for $E(\sigma_{max})$ similar to that of Eq.(14). For a more detailed discussion on the extreme value theory see Lutes and Sarkani (1997). The Vanmarcke and Der Kiureghian formulas are a bit more complex than their Poisson counterparts and are not considered in this study. Prior investigation (Lutes and Sarkani 1996) has shown that for fatigue damage accumulation the difference from using Poisson instead of the Vanmarcke pproximation is minimal although further investigation is needed to determine the significance of bandwidth on the residual stress decay. In this investigation only the Poisson approximation of the extreme value is considered.

Strictly speaking, the expected value of the correction factor $E(CF_j)$ should be evaluated from its probability distribution, which can be found from Eqs.(13) and

(14). A crude but simple approximation for the expected value of a nonlinear function of a random variable is the function evaluated at the expected value of the random variable (Benjamin and Cornell 1970). Following this procedure, $E(CF_j)$ can be approximated as

$$E(CF_j) \approx \left[1 - \left(\frac{\sigma_y - E(\sigma_{max})}{\sigma_{ult}}\right)^2\right]^{-m} \quad \text{for} \quad E(\sigma_{max}) < \sigma_y \tag{15}$$

$$E(CF_j) = 1 \quad \text{for} \quad E(\sigma_{max}) \geq \sigma_y$$

This equation is particularly easy to apply when Eq.(14) is used to compute $E(\sigma_{max})$ This approximation is shown in Fig. 5 for different values of σ_{rms}. As can be expected, the correction factor decays to unity and the rate of this decay depends on the severity of the stochastic loading, as measured by its RMS value.

The use of Eq.(14) also allows a more computationally efficient evaluation of the summation on $E(CF_j)$ and thus of the expected fatigue life. The method adopted here involves approximating the summation by a numerical integral, which yields the following nonlinear equation for the fatigue life N:

$$\int_1^N E(CF)dt = K\left[2^{3m/2}\sigma_{rms}^m \Gamma\left(1 + \frac{m}{2}\right)\right]^{-1} \tag{16}$$

Note, however, that in order to avoid the singularity of Eq.(14) for $n(t)=1$, the lower limit of the integral in Eq.(16) should be taken slightly greater than unity.

Effect of Residual Stress on Fatigue Life

Linear regression on constant-amplitude test data from the T-weld specimens shown in Fig. 1 resulted in an S-N curve in terms of nominal stress range of $N=166{,}000(S/\sigma_y)^{-3.300}$. Taking into account the residual stress model described by Eq.(12) yields the S-N curve in terms of the effective stress range as $N=270{,}000(S_{eff}/\sigma_y)^{-4.039}$. The two curves are in good agreement for $S\approx 2\sigma_y$ at which value the residual stress is removed from the specimen according to the previously presented model. For smaller stress levels the difference is substantial. The change of m from 3.300 to 4.039 shows that fatigue lives of these joints can be increased if the residual stresses are reduced or removed completely. Subjecting the specimen to a precycle in which the nominal stress is equal to the yield stress can effectively do such residual stress removal.

The expected fatigue lives under narrowband loadings with different σ_{rms} levels with and without considering the effect of residual stress decay are presented in Table 1. Both calculations are obtained through Eq.(16). When the residual stresses are neglected, the left-hand side of Eq.(16) becomes exactly the expected fatigue life. When the residual stresses are considered, Eq.(16) is solved using the nonlinear equation solver of Mathcad 6.0. From the numbers in Table 1 it is observed that neglecting residual stress can give predictions of fatigue life that are only 50% of the predictions obtained including the Gerber correction for mean stress. The magnitude of this discrepancy decreases as σ_{rms}/σ_y increases, but large values of σ_{rms}/σ_y are generally unacceptable in practice because of the small magnitude of expected fatigue life. In practice, with the value of σ_{rms}/σ_y being often less than 0.1, it appears that residual stress should be expected to have a significant effect on stochastic fatigue.

Effect of Material Hardening on Decay of Residual Stress and Fatigue Life Prediction

As already pointed out, in the case of a hardening material, the local stress response at the weld toe depends on the elastic-plastic response of the entire cross-section. As a result, it is practically impossible to define a simple analytical model of residual stress decay in the form of Eqs.(15) and (16). The most direct approach in this case is to obtain *E(CF)* using Monte-Carlo simulations. To this end, ten narrowband time histories of the nominal stress $\sigma(t)$ having $\sigma_{rms}=0.25\sigma_y$ have been simulated. Each of these samples contains 10000 cycles. Subsequent analysis confirms that these 10000 cycles are sufficient to completely eliminate the residual stress effects, i.e, to reduce the value of the correction factor to unity. The relatively high RMS value results in the maximum of the loading slightly exceeding the yield stress of the material within the 10000 cycles considered. A segment of one of the simulated loading histories is shown on Fig. 6. The corresponding nominal bending moment acting on the joint is computed as $M(t)=\sigma(t)W$, where W is the elastic modulus of the cross-section. For each simulated loading history, $M(t)$ is applied to the critical cross-section having the initial residual stress distribution shown in Fig.3.

The bending loading is applied to the cross-section in time steps t_k, k=0,1,2.... The stress distribution at the t_k time step is computed by satisfying the following nonlinear set of equations:

-strain increments $\Delta\varepsilon_k(y)=\Delta\theta_k y$, in which $\Delta\theta_k$ is the rotational increment of the cross-section;

-strains $\varepsilon_k(y)=\varepsilon_{k-1}(y)+\Delta\varepsilon_k(y)$;

-stresses $\sigma_k(y)=\sigma[\varepsilon_k(y)]$, in which $\sigma(\varepsilon)$ is the material stress-strain curve taking into account the material's loading history;

-stress distribution must satisfy $M_k = \int y\sigma_k(y)dy$;

A stress response history at the weld toe obtained by this procedure is shown in Fig. 7. Also shown in this figure are the nominal loading $\sigma(t)$ and the response at the weld toe location when the material is elastic perfectly plastic (i.e., there is no hardening). It can be seen from these plots that the mean stresses in a hardening weld joint are slightly higher than those in the elastic-perfectly-plastic case but in very good agreement in general. This is so because under the loading considered, gross yielding of the specimen does not take place – the response of the specimen is elastic as a whole. Under such conditions one may assume that the weld toe location is loaded under "strain control," i.e., under strain $\sigma(t)/E$ rather than under stress $\sigma(t)$. Obtaining the response of a single point of elastic-plastic material with linear hardening yields a stress time-history virtually identical with that obtained at the weld toe considering the whole cross-section.

Because the analysis of a single point is much less computationally involved than analysis of the whole cross-section, the stress time history at the weld toe to a simulated nominal loading are obtained by the former method. Repeating the analysis for each of the ten nominal loadings considered yields ten mean stress time histories at the weld toe. The average of these mean stress histories is shown in Fig.8. The *E*(*CF*) is obtained by first computing the correction factors for each mean stress history using Eq.(8) and then averaging across the ensemble. The average *E*(*CF*) values are shown in Fig. 9. Solution of Eq.(11) yields the expected fatigue life as 5.37×10^5. It can be seen that this life is slightly lower than 5.38×10^5 which was the fatigue life predicted by the residual stress decay model based on elastic-perfectly-plastic material.

Thus, incorporating linear kinematic hardening material model into the stochastic fatigue modeling leads to residual stress decay rates that may be slightly lower than those predicted by the elastic-perfectly plastic model. In general, the decay rates were very close to those predicted by the simple model assuming perfect plasticity. As a result, the values of the Gerber correction factor are slightly higher and the predicted fatigue life is slightly lower. However, the difference in the two predicted lives was less than 1000 cycles (0.2%). Thus, when the overall response of the specimen is elastic and when the effect of triaxial stress fields is ignored, the simple elastic-perfectly plastic model may yield quite accurate results. It is recognized, though, that further studies (different specimen geometries and loadings) are required to establish the range of applicability and the limitations of the presented residual stress model.

Summary and Conclusions

Today's state-of-the-art of nondestructive evaluation of residual stresses in welded joints is based on two diffraction techniques: the X-ray diffraction and neutron diffraction. There are commercially available X-ray diffraction instruments for both laboratory and field testing. Neutron diffraction, on the other hand, is possible only in the presence of a strong neutron source, which is usually found in nuclear reactors. The X-ray diffraction method, like most nondestructive methods for residual stress determination, is capable of measuring residual stresses that are on the surface of a component or structure. Due to the high penetration potential of neutrons, the neutron diffraction method is the only nondestructive technique that is able to provide a complete, through-thickness distribution of the residual stresses in welded joints.

The information on residual stresses is particularly important for fatigue damage calculation under stochastic loadings. A T-welded joint has been investigated and the results show that welding residual stresses can significantly alter the damage accumulation rates. It is demonstrated that the influence of residual stresses on stochastic fatigue damage accumulation may be incorporated by a simple approach based on an elastic-perfectly-plastic material model and the Gerber correction factor. The model assumes that the remaining residual stress at the critical location depends on the largest nominal stress ever seen by the welded joint. The calculations predict that the residual stresses during stochastic loading randomly decay to zero. The effect of material yielding is additionally investigated by considering an elastic-plastic material model with linear kinematic hardening. The residual stresses in this case are computed through Monte-Carlo simulations. Strain hardening must reduce the rate of decay of the residual stress, so must decrease the fatigue life. The very small (0.2%) change in the example is consistent with this assertion, but is not the basis for the conclusion that strain hardening leads to less rapid decay of residual stresses. It can be seen that the effect of material hardening is to reduce slightly the rate of residual stress decay and thus to accelerate the rate of fatigue damage accumulation. For the particular case considered, the predicted fatigue life considering material hardening was only slightly lower than that predicted by the simple elastic-perfectly plastic model.

References

Benjamin, J.R., and Cornell, C.A., 1970, *Probability, Statistics, and Decision for Civil Engineers, McGraw-Hill*, New York, N.Y.

Bishop, N.W.M., and Sheratt, F., 1990, "A Theoretical Solution for the Estimation of Rainflow Ranges from Power Spectral Density Data," *Fatigue Fract. Engng. Mater. Struct.* v. 12, no. 4.

Blom, A.F., 1995,"Spectral Fatigue Behavior of Welded Joints," Int. J. Fatigue, v. 17, no. 7, pp. 485-491.

Davenport, A. G., 1964," Note on the Distribution of the Largest Value of a Random Function with Application to Gust Loading," *Proc. Instn. of Civ. Engrs.*, v.28, pp. 187-196.

Der Kiureghian, A., 1980, "Structural Response to Stationary Excitation," *J. Engrg. Mech. Div.*, ASCE, v. 106, no. 6, pp. 1195-1213.

Fuch, H.O, and Stephens, R.I, 1980, *Metal Fatigue in Engineering*, John Wiley, New York, N.Y.

Goldak, J., 1989, "Modeling Thermal Stresses and Distortions in Welds," in *Recent Trends in Welding Science & Technology, Proc 2nd Intl. Conf on Trends in Welding Research*, pp. 71-82.

Holden, T.M., 1995, "Determination of Macrostresses (Type-I stresses) and Microstresses (Type-II stresses) by Neutron Diffraction," *Proc. Material Research Symposium*, v. 376, pp. 385-390.

Holden, T.M., Root, J.H., Holt, R.A., and Hayashi, M., 1995, "Neutron Diffraction Measurement of Stress," Physics B: Condensed Matter, v.213-214, pp. 793-796.

Holden, T.M., Root, J.H., Holt, R.A., and Roy,G., 1988, "Neutron Diffraction Measurements of the Residual Stress State of a Tubular T-joint," *Proceedings of the 7th International OMAE Symposium*, Houston, Texas, pp.127-131.

Jo, J., Wang, X., L., Kleinosky, M.J., Green, R.S., Hubbard, C.R., and Spooner,S., 1993, "Evaluation of Stress Relief Treatments Using Diffraction," *Proc. 4th International Conference on Residual Stresses*, eds., S.A. David and J.M. Vitek, ASM International, Materials Park, OH.

Kihl, D.P., and Sarkani, S., 1996, "Mean Stress Effects in Steel Weldments Subjected to Constant Amplitude Fatigue Loadings," *Proc. OMAE'96.*

Klassen, R.J., Root, J.H., and Holden, T.M., 1991, " Neutron Diffraction for Non-destructive Evauatoin of Internal Stress in Engineering Components," Proc. 37th International ISA Symposium, v.37, pp.697-706.

Lutes, L.D., and Sarkani, S., 1996, "Decay of Residual Stress in Stochastic Fatigue," *J. Struct. Engng.*, ASCE, v.122, no.1, pp.92-98.

Lutes, L.D., and Larsen, C.E., 1990, "Improved Spectral Method for Variable Amplitude Fatigue," *J. Struct. Engng.*, ASCE, v. 116, no. 4, pp. 1149-1164.

Lutes, L.D., and Sarkani, S., 1997, *Stochastic Analysis of Structural and Mechanical Vibrations*, Prentice Hall, New Jersey.

Mahin, K.W., Winterss, W.S., Holden, T.M., and Root, J.H., 1991, "Measurement and Prediction of Residual Elastic Strain Distribution in Stationary and Traveling Gas Tungsten Arc Welds, in *Practical Application of Residual Stress Technology*, ed. C.O.Ruud, ASM International, Materials Park, OH.

Miner, M.A., 1945,"Cmulative Damage in Fatigue," *J. Appl. Mech.*, 12, A159-A164.

Palmgren, A., 1924, "Die lebensdauer von kugellagen," *Ver. Deut. Ingr.*, 12, pp. 339-341 (in German).

Sarkani, S., and Lutes, L.D., 1988, "Residual Stress effects in Fatigue of Welded Joints," *J. Struct. Engng.*, ASCE, v.114, no.2, pp. 462-474.

Vanmarcke, E.H., 1975, "On the Distribution of First-Passage Time for Normal Stationary Random Processes," *J. Appl. Mech.*, 42, pp. 215-220.

Table 1 Predicted Stochastic Fatigue Lives

$\frac{\sigma_{rms}}{\sigma_y}$	Predicted Fatigue Lives	
	Residual stress decay neglected	Residual stress decay included
0.02	1.46×10^9	3.16×10^9
0.025	7.00×10^8	1.45×10^9
0.035	2.31×10^8	4.63×10^8
0.05	7.11×10^7	1.44×10^8
0.06	3.89×10^7	7.97×10^7
0.075	1.86×10^7	3.94×10^7
0.10	7.21×10^6	1.57×10^7
0.15	1.89×10^6	3.99×10^6
0.20	7.33×10^5	1.32×10^6
0.25	3.51×10^5	5.38×10^5
0.33	1.40×10^5	1.75×10^5

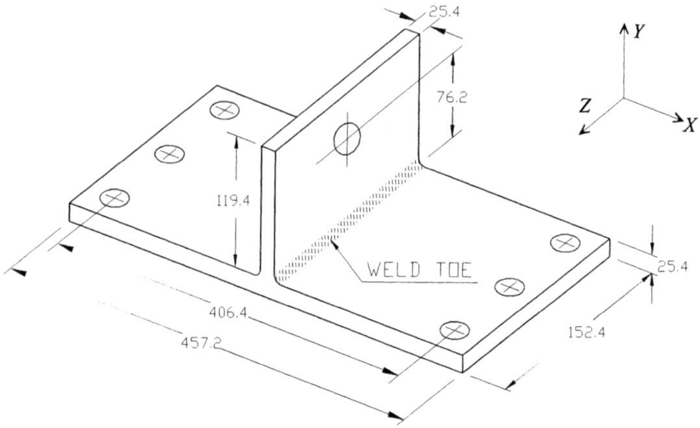

Fig. 1. Fatigue Specimen

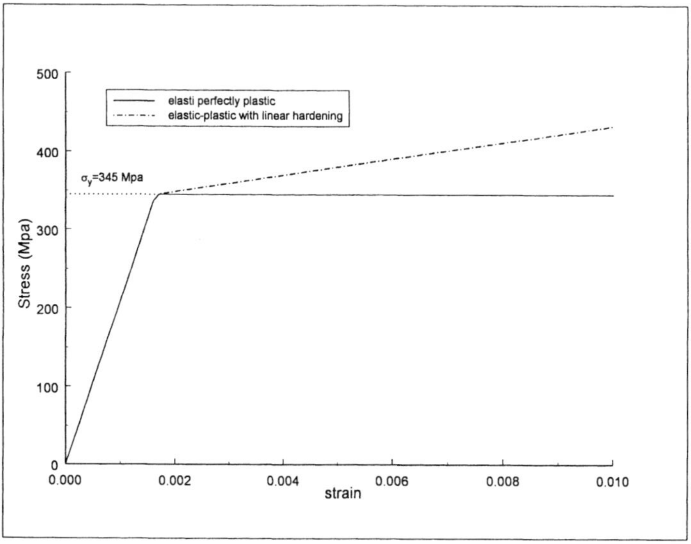

Fig. 2 Material models used in the investigation

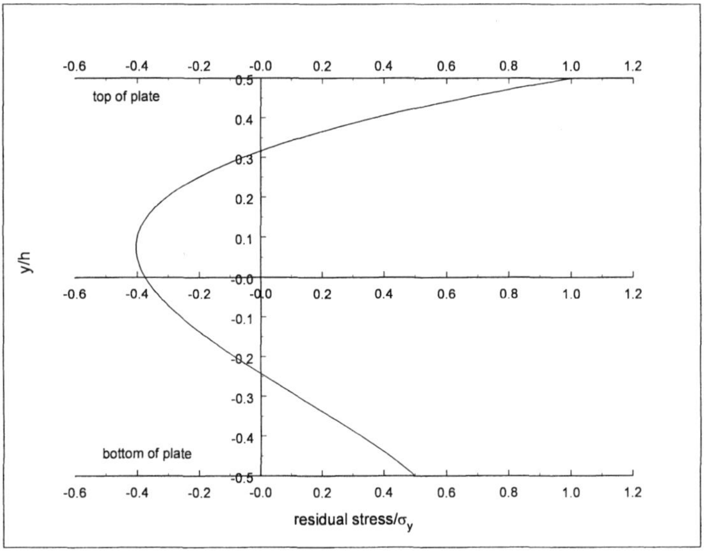

Fig.3. Assumed residual stresses due to welding at cross-section of weld toe

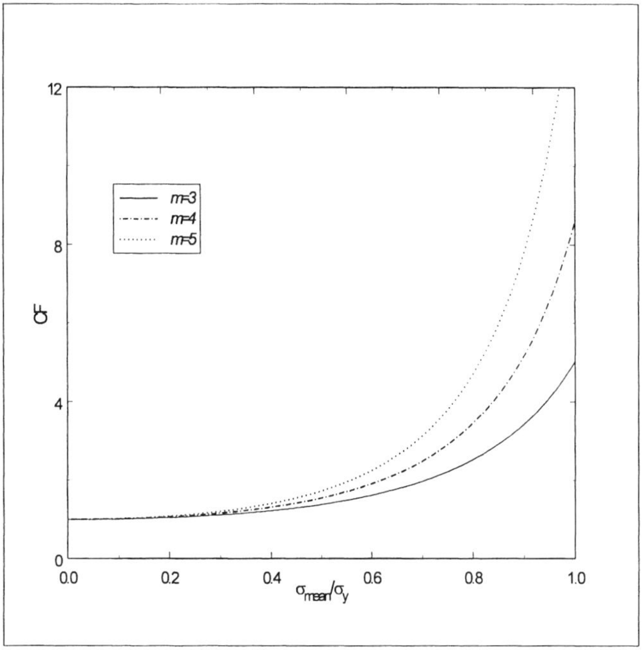

Fig. 4. Gerber correction factor

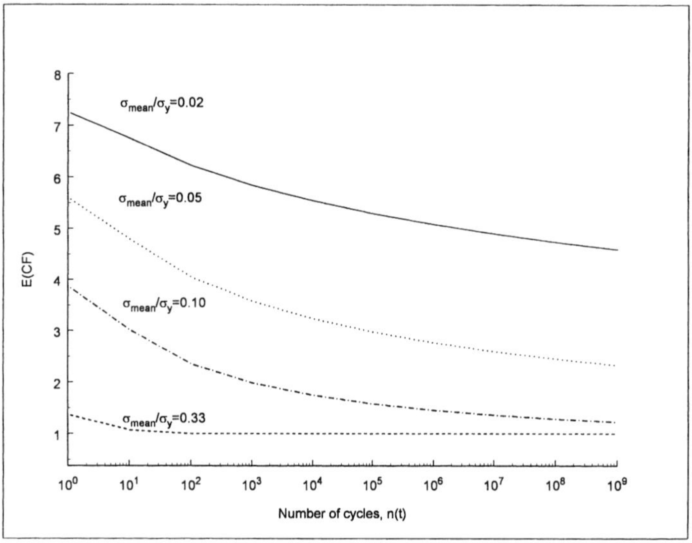

Fig. 5. Expected value of Gerber correction factor

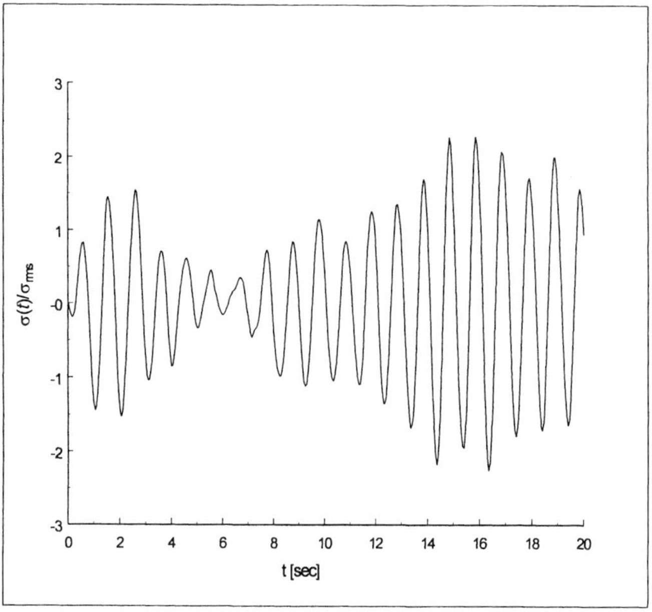

Fig. 6. Segment of one of the simulated nominal stress histories

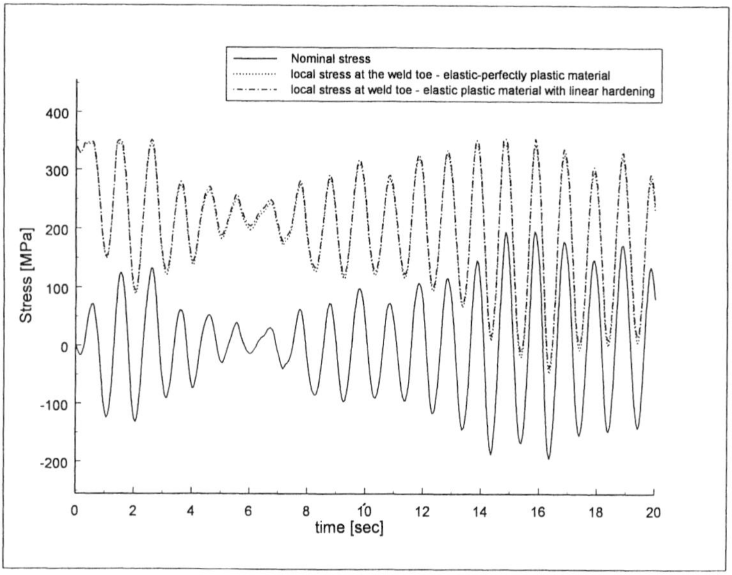

Fig. 7. Local responses at the weld toe

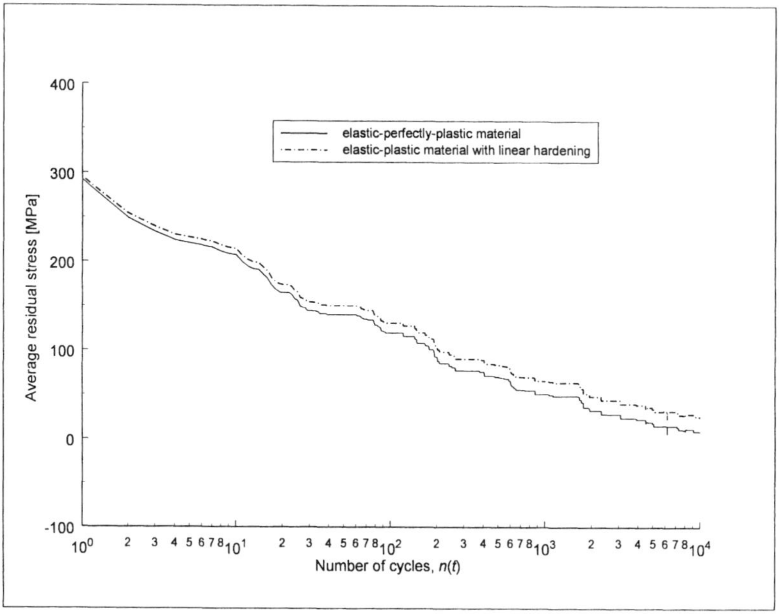

Fig. 8. Average decay of mean stresses

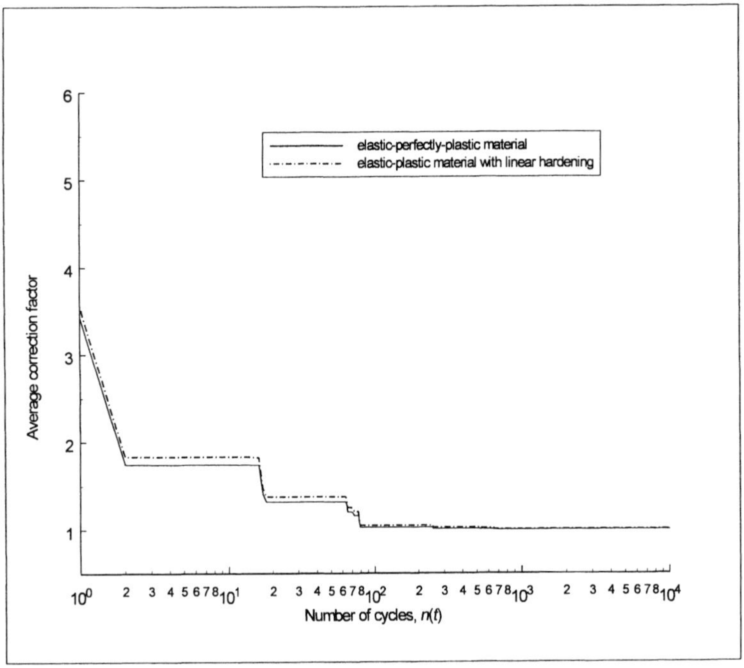

Fig. 9. Average Gerber correction factor

Finite Element Simulation of Welding Residual Stresses

Shahram Sarkani[1] and George Michaelov[2]

Abstract

Prediction of residual stresses in welded joints using nondestructive test (NDT) methods such as X-ray or neutron diffraction techniques can often benefit from numerical analysis of those stresses. Such analyses, which can be carried out either before or after the NDT procedure, are most reliably done with the help of powerful finite element (FE) simulation techniques. Although the finite element method is one of the most attractive approaches for computing residual stresses in welded joints, its application to practical analysis and design problems has been hampered by computational difficulties. These difficulties do not arise in modeling the complex constitutive response of melting and solidifying metal; rather, they occur mostly because of the enormous computational size of any practical problem resulting primarily from the three-dimensional (3D) modeling of a welding process. Although two-dimensional (2D) modeling has been used widely in residual stress problems, current belief holds that 2D analysis cannot render accurate residual stresses that occur due to welding. This study investigates the residual stress fields in a welded T-joint, comparing those computed by 3D models to those computed by 2D models. The study shows that the temperature distribution in the central zone of the joint can be captured successfully by a 2D finite element model and a technique that takes into account the heat transfer balance and welding speed. The residual stresses in the plane of the 2D model computed by this method show fairly good agreement with those computed by the 3D model.

[1] School of Engineering and Applied Science, The George Washington University, Washington, DC 20052
[2] Pragma Systems Corporation, Reston, VA 20190

Introduction

Residual stresses in a welded joint occur throughout the area of solidified weld metal and the heat-affected zone (HAZ). The mechanism by which these stresses are formed is well known. Due to the high intensity of the localized heat source, a nonuniform transient temperature field occurs. This field initially causes a rapid thermal expansion followed by a thermal contraction of the affected material. The severity of the induced temperature gradients and the degree of restraint the joint imposes on the thermal deformations determine the residual state of the weldment when it has finally cooled. This residual state is characterized by a combination of internally balanced residual stresses and weld distortion.

From the perspective of continuum mechanics, the welding process can be viewed as a transient boundary value problem. The constitutive equations in this problem take into account the physics of heat transfer and the mechanics of thermal dilatation, as well as the processes of change in material microstructure and phase transformation. The boundary conditions model the welding heat input, the surface heat losses, the mechanical restraints, and, most importantly, the contact between the welded parts and the infill metal deposit. Practical solution of such a complex boundary value problem became possible only with the computer implementation of powerful numerical techniques, among which the finite element method (FEM) emerged as the most powerful. A number of studies have investigated the application of FEM to welding residual stress modeling. The complexity of the FE residual stress models used, however, has kept most fatigue researchers and design practitioners to the experimental nondestructive testing (NDT) techniques for directly measuring residual stresses. The most popular NDT methods include the X-ray diffraction method (Prevey, 1986), the neutron diffraction method (Holden, 1995; Holden et al., 1995, 1996), and the holographic interferometry method (Nelson, et al., 1994), just to name the most popular. All NDT methods, though, have two common disadvantages. First, they are expensive to apply because their application usually requires special equipment and personnel. Second, usually residual stresses are measured at discrete locations on a weld, and close to the weld surface. A complete picture of the residual stress distribution in a general weldment is difficult and expensive to obtain by experimental techniques alone. It is an important question, therefore, whether simplified computationally efficient FE models can be developed which can be used to supplement and enhance an experimental NDT investigation of residual stresses. A preliminary numerical analysis of welding residual stresses can point out the optimum locations for the experimental setup and can be useful in relating the residual stress values at the measured discrete locations to the entire distribution of the residual stress field. Note that residual stresses cannot cause weld failure directly. Plastic deformations and the resulting stress distribution are a very good mechanism by which metals avoid cracking due to localized stresses above the yield point. Initial distribution and subsequent redistribution of residual stresses are important for fatigue life calculations (see Sarkani, Michaelov, and Lutes (to appear). Development of simplified FE technique for approximate computation of welding residual stresses is the primary objective of this investigation.

It should be noted that finite element modeling has already been found particularly useful in the application of the X-ray diffraction method to measure through-thickness residual stress distribution. In this technique, the FE method is used to provide for correction to the stress relaxation occurring as a result of the material removed from the exposed layer. Such correction procedures were first reported by Pedersen and Hansson (1989) but were more rigorously investigated by Lambda Research in the 90s (Lambda Research, 1996). The FE analyses in these procedures are relatively simple because they do not require direct simulation of the residual stress fields. Although this FE correction technique allows a more thorough determination of the thickness residual stress gradients with the X-ray diffraction method (which is not possible by the X-ray method alone) the whole technique becomes destructive. If the measurements are to be made on a real structure rather than on a specimen, it may be preferable to limit them to the surface residual stresses only, while investigating their distribution through the weld by the FE simulation approach.

Most FE investigations of residual stresses take into account that the thermo-mechanical processes that occur when the weld solidifies evolve naturally along the travel direction of the weld torch. At the same time, heat dissipation and mechanical constraints on thermal dilatations are influenced by the thickness and geometry of the plate joints; that is, by conditions in a direction transverse to the welding direction. This makes a welded joint truly three-dimensional (3D) and calls for a corresponding 3D finite element simulation. Such an analysis, although relatively straightforward as a theory, is difficult to perform in practical analysis and design due to a significant computational demand. A 3D FE simulation of welding is computationally demanding because the mesh it requires in order to compute accurately the temperature and stress quantities is relatively refined due to the high temperature gradients. Also, the nonlinear material behavior due to plasticity and temperature dependence of properties requires small time increments to achieve convergence. Furthermore, many welds are created by multiple passes which, when reflected in the FE analysis, can increase the computational time even more. The computational demand of the 3D models is so prohibitive that all FE investigations of residual stresses in welds before the late 1980's were performed on simplified two-dimensional (2D) models. Note that for non-symmetric and near edge cases one has no choice but to model the 3D nature of the problem.

In 2D models, only a plane perpendicular to the direction of the weld is considered. The behavior out of plane can be taken as plane stress (assuming zero out-of-plane stresses), plane strain (assuming zero out-of-plane strains), generalized plane strain (assuming constant strain normal to the model plane), or axysimmetric. Among the investigations considering such 2D models one should mention those of Hibbitt and Marcal (1973); Friedman (1975); Anderson (1978); Rybicki, et al. (1978), Rybicki; and Stonesifier (1979); Karlsson (1986); and Mahin, et al. (1988). One can also classify the shell models (Lindgren 1988) as 2D. As is evident from the literature, 2D modeling was used as a basis for the development of weld simulation techniques. Since investigators tried to avoid modeling in 3D, the computed residual stresses were verified by comparison with experimental measurements. Certain discrepancies between computed and experimentally measured residual stresses were reported. This led to the belief that 2D models are inadequate to quantify the residual stresses accurately in a welded joint. Two points should be made here. First, some of the error in computed residual stresses should be attributed to an incomplete understanding of the physics of solidifying weld metal and to the difficulties in obtaining high-temperature material constitutive data (Goldak, 1989). Second, for the 2D-model plane, fairly good agreement between computed and measured stresses is reported in a number of studies [e.g., Rybicki, et al. (1978); Rybicki and Stonesifier (1979); Mahin, et al. (1988)]. The largest discrepancies were found for out-of-plane stresses such as the hoop stress in axysimmetric models. "Do plane stress and plane strain solutions bound the 3-D solution?"

Three-dimensional modeling of welds was first attempted by Tekriwal and Mazumder (1988, 1991) and, independently, by Karlsson and Josefson (1990). Tekriwal and Mazumder performed a simulation of a single-pass butt weld formed of two 5.8-mm thick plates. Karlson and Josefson analyzed a single-pass weld of an 8-mm thick pipe with diameter equal to 114.3 mm. These analyses confirmed the 3D nature of the temperature and stress fields developing during welding, but at the same time demonstrated the restrictions of the 3D calculations. The investigators were limited to relatively coarse FE meshes to accommodate the analysis in the computer facilities available to them. The dimension along the weld of Tekriwal's model, for example, was only 25.4 mm–hardly representative of a real weld. In the following years, more studies of welding residual stresses using 3D FE modeling of welds were reported [e.g., Gu and Goldak (1994); Goldak, et al., (1996); Oddy and McDill (1996)]. These later studies, however, do not report analyses that are quantitatively different from those conducted in the late 1980s. This is because the speed of the commonplace engineering workstations has roughly increased only by a factor of 10 over the last decade, while halving the FE mesh size in welding simulation could increase the computational cost by a factor of 100 (Goldak, 1989).

Significant refinement of a 3D FE mesh, modeling of long and multi-pass welds are likely to remain impractical tasks in the foreseeable future for all but the most powerful computers. A question therefore arises about whether a 2D model may be "tuned" so as to provide a satisfactory match to the results obtained by a general 3D model. Such a "tuned" 2D model may provide an invaluable tool to designers of welded metal structures, fatigue analysts, and other welding specialists. Consistent analysis of this problem is currently lacking. In fact, comparisons between the results of corresponding 3D and 2D simulations of the same weld problem are practically not existent. Although the temperature and the stress fields in a weld formed under a moving heat source are transient and three-dimensional, some authors [e.g., Karlsson and Josefson (1990)] report the development of quasi-stationary conditions away from the weld ends, particularly in the temperature fields. A properly constructed 2D model should capture this quasi-stationary behavior at desired positions of the weld.

The objective of this study is to investigate the possibility of "tuning" a 2D-weld model onto a corresponding 3D-weld model for the purpose of calculating the magnitude and distribution of the welding residual stresses. To this end, 3D finite-element simulations of a single and a three-pass T-weld are carried out. It will be demonstrated that a simple technique of matching the total thermal input during welding can be used in connection with a 2D heat transfer model to provide a relatively accurate picture of the transient temperature fields. These 2D temperature fields are used in a corresponding 2D mechanical model to compute the welding residual stresses. A comparison between the results of the 2D and the 3D models is carried out. The ABAQUS/Standard finite element code is used throughout the investigation.

Constitutive Relations and FE Modeling of Welding Residual Stresses

It has become common in FE welding simulation to perform heat transfer and thermal stress analysis in two consecutive steps. Although the whole welding process can be simulated in one step by a coupled thermo-mechanical analysis, the uncoupled technique allows all computations to be performed faster and to utilize fewer computer resources. The results of the uncoupled and coupled analyses are usually very close since the mechanical deformation rates are small enough not to influence the temperature fields.

Accordingly, for the first step – the transient heat transfer – the constitutive relations are given by the following diffusion equation:

$$\rho C_p \dot{T} = \underline{\nabla} \bullet k \underline{\nabla} T \qquad (1)$$

in which ρ is the material density, C_p is the specific heat capacity of the material, k is the material thermal conductivity, T is the temperature, $\dot{T}$ is the temperature rate, and $\underline{\nabla} = \underline{i}_1 \frac{\partial}{\partial x_1} + \underline{i}_2 \frac{\partial}{\partial x_2} + \underline{i}_3 \frac{\partial}{\partial x_3}$ is a 3D gradient vector operator. The heat transfer boundary conditions of the problem are

$$q = -k \underline{\nabla} T \qquad (2)$$

in which q is the heat flux at the boundary. In welding problems, the heat flux q may consist of one or more of the following modes: convective heat loss, radiative heat loss, prescribed heat flux across the boundary, and zero flux in a plane of symmetry.

The temperatures histories computed during the heat transfer step at the nodes of the FE mesh are calculated and saved. During the second step of welding simulation, these saved temperature histories are used as the thermal loading to produce the residual stress field. Each increment of the analysis assures that stress equilibrium is satisfied through the weld and at the boundaries (*ABAQUS 5.7 Theory Manual*, 1997). The material is assumed to obey elastic-plastic constitutive relations with a von Mises yield function, associated flow rule, and isotropic hardening. It should be noted that some authors (e.g., Tekriwal and Mazumder, 1991, assert that the kinematic hardening is more appropriate than isotropic hardening in a welding simulation. Isotropic hardening is assumed in this study for simplicity and to reduce the computation cost; moreover, the main objective here is to compare the results of 3D and 2D modeling. The material is completely characterized by its Young's modulus, Poisson ratio, and uniaxial yield stress-plastic strain curve. All these characteristics are assumed to be temperature-dependent.

In addition to the regular FEM procedures for solving transient nonlinear problems, the approach to residual stress determination requires that the depositing of melted weld metal be simulated. This simulation is required twice: once in the heat transfer and once in the stress analysis. In ABAQUS, welding is most conveniently simulated by the gradual addition (or activation) of finite elements based on assumed weld speed. The thermal input due to welding is modeled by assigning body fluxes to the added elements and surface fluxes to the adjacent plate elements in the contact zone. During the heat transfer step, ABAQUS contact modeling techniques are used to specify the thermal interaction between the newly added elements and the part of the weldment that has already been formed. During the stress analysis step, the newly added elements are directly constrained to the weldment in a stress-free state.

The analysis described above can be applied to any weldment configuration. The design of the FE mesh, however, needs to address certain issues so the analysis is carried out efficiently and the results obtained are as accurate as possible. In particular, proper consideration should be given to modeling the heat-affected zone (HAZ). The highest temperature gradients usually take place in the HAZ in planes perpendicular to the welding direction. As a result, the temperature dependent material properties may change significantly in the HAZ over relatively small distances. All this requires a relatively refined FE mesh. The mesh in the HAZ should also be refined in the direction parallel to the welding. This is necessary so the discontinuous simulation by element addition matches the continuous infill metal deposition process as close as possible. These elements can be added in groups, or layers, to simplify the procedure; however, the thickness of these layers should be small enough to achieve a satisfactorily accurate solution. The aspect ratio (i.e., the ratio of the largest to the smallest dimension) of the 3D elements also restricts the thickness of the layers. Use of thinner element layers, however, increases the number of analysis steps needed to model the entire welding process. As a result of all these refinements, a FE model of significant size may be created, and the time required to run the analysis of such a model may be prohibitively large.

The simplest modeling feature of which an analysis can take advantage is model symmetry, which allows only half of the weldment to be included in the FE mesh. This advantage should be used with caution, however, because in certain cases the finished weldment may be symmetric while the welding process may not be. In a T-weld, for example, symmetry implies that weld material is placed on both sides of the weld at the same time. Currently, no data exist to show whether or not modeling a nonsymmetric weldment with a symmetric FE model can provide satisfactory results.

One approach to reducing computations is to simplify the material properties. Some investigators (Tekriwal and Mazumder, 1991) "cut" the temperatures above 600°C (1112°F) in the stress-strain analysis in order to reduce the influence of the temperature-dependent plasticity on the number of stiffness matrix re-evaluations during the stress-strain analysis. This simplifying assumption is justified because the elastic modulus of steel is low enough above this temperature, so that the strains acquired when the temperature is above the cut-off point contribute insignificantly to the final residual stress state.

An approach commonly used in modeling multi-pass welds is to simplify the welding process by reducing the number of weld passes by grouping them (Rybicki and Stonesifer, 1979; Mok and Pick, 1990; Leung, 1986). A variation of this technique is to carry out the analysis of each group in two steps (Leung, 1986). During the first step, all the physical passes composing the group except the final pass are applied. The second step considers the final pass completing the group. This approach is especially attractive for applications in which the surface residual stresses (as crack promoters) are of most interest. In such a case the last weld pass can be assumed to have the most significant effect on the surface residual stresses.
Brown and Song (1993) have suggested a FE re-meshing technique at each step of the analysis as a means of reducing computational cost. Accordingly, an extremely fine FE mesh is used only in the "hot" part of the model – the HAZ and the weld. At the same time, in the "cold" part of the model (or the part where the rate of temperature change is much lower) a much more coarse mesh can provide satisfactory accuracy. In fact, the rate at which the temperature changes determines the mesh size. Re-meshing is also a computationally expensive technique because it requires the stress-strain results obtained during a previous analysis step to be mapped onto the current one. It should be noted also that Brown and Song applied the technique to a shell model, which does not take into account weld metal deposit. The technique seems directly applicable to laser welds; however, it certainly has potential for application to arc welds.

The approach that can lead to a most significant reduction of the computations is to approximate a 3D-weld model by a corresponding 2D. This approach is investigated below for the particular case of a T-weld.

3D Finite Element Analysis of a Welded T-Joint

The geometry of the weld considered in this study is shown in Fig. 1. The plates of the joint are 25.4 mm (1") thick and are assumed to have been welded simultaneously on both sides so that the welded joint is completely symmetric and only one-half needs to be modeled. The joint is 76.4 mm (3") long in the welding direction. Both single-pass (Fig. 1a) and three-pass (Fig. 1b) weldments are considered. The properties of the material (base metal and welding) are taken from Terkriwal and Mazumder (1988,1991). These properties should be representative of mild steel (0.22% carbon). It should be noted that these temperature-dependent properties are similar to those used by other investigators (e.g., Rybicki and

Stonesifer, 1979, Karlsson and Josefson, 1990) for other types of steel. Also note that the mechanical properties of steel above certain temperatures (1000°C or 1832°F) may become uncertain, but for numerical reasons they are assigned distinct values here.

Following the procedure previously outlined, the weldment of the three-inch-long T-joint is divided into 15 layers of 5.08 mm (0.2 inch) each along the welding direction. For both the single-pass and the three-pass weld, the welding travel velocity is 114.3 mm/min (4.5 inch/min); thus the full analysis cycle for adding elements is 2.65 seconds per layer. The total effective thermal input for the single-pass weld is assumed to be 9.8 kW (9.3 BTU/sec). This corresponds to an electric current of 350A, a voltage of 35V, and a total arc efficiency of η=0.8. For the three-pass models the effective thermal input is assumed to be or 5.4kW (5.1 BTU/sec) for each welding pass, which corresponds to 250A, 27V, and η=0.8 total arc efficiency. The assumed values of the electric current and voltage are relatively high compared to the average values discussed by various investigators, but this study considers relatively thicker plates (25.4 mm), which require relatively greater thermal input. The assumed value of the total electric arc efficiency η=0.8 lies in the upper region of the relatively broad range adopted by different investigators [from η=0.55 in Legatt, (1991) to η=0.9 in Karlson and Josefson (1990)]. The reason for this assumption is that the geometry of the T-joint and the simultaneous use of two symmetric torches significantly reduce the thermal losses.

Analysis reveals that the temperature time histories along the direction of the welding differ by little within a relatively long central zone. This central zone is situated at a distance of approximately 1.00-1.25 times the weld plate thickness away from the plate edges. Thus, within the central zone, pseudo-stationary conditions (i.e., lack of significant variation along the weld direction) take place in the temperature fields. The existence of such a central zone with pseudo-stationary conditions for the residual stress fields is confirmed by the subsequent stress analysis. Figures 2 and 3 show samples of the S33 residual stresses (those normal to the weld direction) computed at different planes along the length of the joint for the single pass and three-pass welds. Among all stress components the S33 stresses were found to have the largest variation along the weld length, which is why they were chosen for these figures.

The central zone represents the most interesting part of 2D modeling because within this zone, due to the pseudo-stationary conditions, plane strain conditions may be expected to prevail. Currently, there does not seem to be an algorithm that can accurately estimate the conditions for development of such a central zone. Based on the results of this study, the length of the weld can be empirically estimated to be at least three times the thickness of the plates, so that a pseudo-stationary central zone can develop. In the following analysis 2D modeling is used to attempt to match the temperature and the residual stress fields obtained at the center layer of the 3D model.

Modeling Heat Transfer Using a 2D Model

Analysis of the nodal temperature histories obtained with the 3D FE model suggests that for a chosen layer of the 3D FE mesh, the following phases of heat exchange may be distinguished:

- *Pre-heating.* This phase is characterized by a small but increasing thermal input to the layer under consideration. During this process, the welding torch has not yet reached the layer; however, the thermal conductivity of the welded plates is sufficient to cause the layer to heat (heat travels faster than the weld torch). There is heat convection at the boundary surfaces of the layer; however, these thermal losses are generally lower than the positive thermal input from the remote torch.

- *Direct heating.* This phase is characterized by the arrival of the welding torch at the layer under consideration as a result of which the layer receives a massive input of thermal energy. This thermal input takes the form of body heat fluxes through the weld filler metal and additional surface heat flux through the adjacent joint areas. The thermal exchange between the layer under consideration and the layer preceding it on the welding path is relatively small because the two layers have similar high temperatures. A more significant convection with the quite cooler succeeding layer takes place during which the layer under consideration loses certain thermal energy.

- *Post-heating/cooling.* During this phase, depending on the position along the joint of the layer under consideration, the net heat exchange between it and the two adjacent layers can be gain or loss. The end of this phase can be assumed when the temperature gradient across the adjacent layers becomes much lower than the gradients within the layer under consideration. At that moment, a pseudo-stationary condition along the weld direction is achieved and the main heat exchange of the layer is with the environment through its free surface.

- *Total cooling.* This phase is characterized by a low heat exchange between the adjacent layers and heat convection at the boundaries until, the weld reaches the same temperature as the environment surrounding it.

Evidently, a 2D heat-transfer model is most appropriate for modeling phases (2) and (4) because the major part of the heat exchange during these periods is in the x_1-x_2 plane (according to Figure 1). Phase (2) can be simulated directly by the same approach to heat flux modeling as that used in the 3D model, with an adequate reduction to account for the thermal losses due to the heat exchange between the adjacent layers. Phase (4) can be accounted for as long as the 2D model has the same film properties on the free surface as does the 3D. Quantifying the effect of either phase (1) or (3) would require modeling the out-of-plane heat exchange.

In the following part of this section a method is presented that allows the effect of pre-heating [phase (1)] on the temperature time histories of a welded joint to be accounted for by a 2D model. Thus, the 2D model developed in this study can account for heat exchange phases (1), (2), and (4) of the previous classification. Phase (3) is much more difficult to account for by a 2D model and will be neglected in the analysis. Overall, this step may result in a small retardation of cool down of the weld section under consideration. As the results obtained show, its effect on the computed temperatures is second-order compared to the effects of the other three phases. The target of such a 2D modeling is a particular section of the joint that lies perpendicular to the welding direction with a specified thickness. Such a section roughly corresponds to a particular layer of the 3D model.

The method is based on the assumption that the two most important issues in modeling heat transfer during welding are: (1) to reproduce the total thermal input that a target section of the joint receives, and (2) to account for the rate at which the target section receives this thermal input (i.e., to account for the welding speed). Note that in a 3D model these two issues are naturally accounted for because a constant surface heat flux is specified on varying element groups at each analysis step, simulating the travel of the welding torch. In a different way, the 2D model presented below uses the same weld elements but applying varying heat fluxes.

Let Q_t denote the total amount of heat energy imparted to the joint for each position of the welding torch. The modeling approach used in this study assumes that this thermal input can be divided into two components. The first component represents the energy spent on melting the filler metal and raising its temperature. In the FE analysis, this component is modeled by body fluxes prescribed through the elements of the filler metal. The second component represents the energy directly imparted to the welded plates. In the FE analysis, this component is modeled by prescribed surface fluxes on the sides of the FE elements that are within the assumed range of the torch. Evidently, in the 3D model the total thermal input Q_t per weld length is equal to the integral sum of the body and surface fluxes per thickness of a 3D layer. If the objective of the 2D analysis of a weld section is to match the result of a particular 3D analysis, the Q_t assumed in the 2D model should be exactly the same as the thermal input per layer in the 3D analysis.

In the 2D analysis, the rate of the thermal input (controlled by the welding speed) can be accounted for by prescribed analysis steps of the same number and duration as those in a corresponding 3D analysis. For the 2D analysis of a particular joint section, an analysis step belongs to the pre-heating phase described previously if the welding torch is currently at a different section. During such a step, no filler metal is added to the section; however, the section receives a certain part of the remotely released heat energy Q_t because heat "travels" much faster than the welding torch. For each finite element of the 2D model, this energy is modeled as a prescribed body flux Q_i. The exact relationship between Q_t and Q_i is difficult to obtain. In this study, the following approximation is assumed:

$$Q_i(x_1, x_2, \Delta x_3) = \alpha(\Delta x_3)\beta(x_1, x_2)Q_t \qquad (3)$$

Here, α and β are modeling parameters. The parameter α depends on the distance Δx_3 of the weld torch from the joint section being analyzed and is different for each analysis step. It is assumed that α is proportional to $(1/\Delta x_3)^2$ (Tekriwal, 1989; Tekriwal and Mazumder, 1988, 1991). The second parameter β depends on the distance of the element from the heated surface of the plates. The parameter β is assumed to be inversely proportional to the distance from the center of the element to the heated surface of the plates. This coefficient is constant at each analysis step. In order to simplify the solution, the elements of the 2D model are divided into several groups with a constant β for all elements within each group. For elements located at relatively large distances from the heated surfaces β was assumed to be equal to zero.

Once the welding torch arrives at the section under analysis, the direct heating phase takes place. The portion of the energy imparted due to direct heating is modeled in the same manner for the 2D model as it is for the 3D model, i.e., by body fluxes through the filler metal and surface fluxes through the heated surfaces. Additionally, to take into account thermal losses to the sections adjacent to the section under analysis the total heat is reduced by a factor of γ; that is, the total heat energy imparted to the section is assumed to be γQ_t. Accordingly, all applied body and surface fluxes are reduced by a factor of γ. This factor is computed as

$$\gamma = \frac{Q_t - Q_p}{Q_t} \tag{4}$$

in which Q_p is the accumulated thermal input imparted to the specified section during the pre-heating phase. Once the welding torch passes the section under consideration, the direct heating phase is complete and the model is allowed to cool down by thermal exchange with the environment through the free surfaces of the 2D model.

The analytical results presented hereafter are for the section in the middle of the 76.2 mm long joint having the same thickness as the 3D model layers. Thus, the 2D model targets the temperature history of the middle (eighth) layer of the 3D model. A comparison of the temperature histories obtained by the 2D approach presented here and the time histories of the eighth layer of the 3D model reveals that the results of the two analyses are fairly close. For the welded plate locations under consideration, which are close to the plate surface in direct contact with the filler metal, the 3D temperature histories can be seen to have two peaks, while those obtained with the 2D analysis have a single peak. This is the biggest difference between the obtained 2D and 3D temperature histories. It should be noted, though, that the double peak of the 3D temperature plots is not a physical feature; rather, it is a FE modeling phenomenon. The FE simulates the filler metal deposition in the form of sudden additions of groups of elements for each model layer. When a location is in close proximity to two layers, some cooling always occurs between the moments when the elements of the two layers are added. In this respect, the results of the 2D model should be taken as being a better representation of reality. Furthermore, this mismatch in the temperature histories should not be important for the comparison of the resulting residual stresses because, as previously mentioned, high temperatures do not contribute significantly to the ultimate residual stress state.

The 2D weld models considered in this study use generalized plane strain elements. A mesh of generalized plane strain elements is similar to a single layer of three-dimensional elements between two rigid planes. The elements of the mesh are defined like conventional plane strain elements except that they all share two extra nodes, which are the reference nodes controlling the relative motion of the rigid planes. Unlike a conventional 2D plain strain a generalized plane strain model allows for some thermal expansion in the model's out-of-plane direction, which results in a better approximation of the 3D situations. Unlike a real 3D model, however, the out-of-plane thermal expansions of the elements of the 2D generalized plane strain model are not independent of each other but are controlled by the motion of the two rigid planes.

A generalized plane strain model is used to compute the residual stresses under two sets of temperature time histories. The first set contains the temperature histories obtained at the central (eighth) layer of the 3D model. The second set uses the temperature histories obtained with the newly developed 2D heat transfer technique. A comparison of the residual stresses obtained with the two temperature sets is shown in Figures 4 and 5 for the S11 and S33 components of the stress tensor. The two result sets are in fairly good agreement, both qualitatively and quantitatively. These results demonstrate that the 3D nature of the temperature fields in a welded joint can be accounted for by the modeling technique developed in this study. Therefore, any differences between the residual stresses computed with a 2D model and those computed with a 3D mechanical model can be attributed to the 2D model's inability to account for the out-of-plane constraints, but cannot be attributed to inaccuracies of the temperature fields.

Computation of Welding Residual Stresses with 2D and 3D models

The 2D technique described previously for computing welding temperature histories can be applied to any section of a welded joint as long as this section is within the "central zone," as defined above. Since the temperature histories are different at each section, the residual stresses computed from these temperatures will also differ. Thus, the new technique allows for differentiation between residual stress fields at different joint sections. The degree to which the out-of-plane constraints of the 2D mechanical model can adequately match the constraints of the 3D model will ultimately determine the efficiency of the 2D model. To consider this issue, the residual stresses computed at the center of the 3D welded joint are compared to those obtained with a 2D mechanical model using the same temperature histories. The 2D models analyzed here included plane stress, plane strain, and generalized plane strain. The generalized plane strain model gave the best match to the 3D stress results and its results are presented below.

Figures 6 and 7 represent a comparison of the residual stresses obtained with the 3D and 2D stress-strain models for a single-pass weld, and Figures 8 and 9 show the results from the three-pass weld. For both weld cases fairly good agreement both qualitatively and quantitatively can be seen between the in-plane stresses S11 from the 2D and the 3D models. (Similar agreement can be seen to exist for the other in-plane stresses, S22 and S12, which are not presented in this article.) A more significant difference (about 30% in the maximum and minimum values) is obtained in the out-of-plane stresses S33 (Figures 7 and 9).

Note that the FE technique of weld modeling by element addition is not a continuous process. It can be observed that the S33 stresses in the 3D model are much more sensitive to this discontinuous modeling than are the S11 stresses. Since the welding is a continuous process, the discontinuities obtained with the 3D model do not represent a physical phenomenon. This makes the S33 results obtained with the 3D model less reliable.

The relatively larger difference in the S33 stresses obtained with the 2D and 3D models is due mainly to the difference in the deformation fields in the x_3 (out-of-plane) direction. The out-of-plane deflections, particularly in the weld filler metal, cannot be captured accurately by any 2D element. This deformation may eventually be captured by a “local” 3D model; that is, by a model built with 3D elements but, representative of a narrow section of the weld and not of the entire welded joint. On such a local 3D model different out-of-plane boundary conditions may be applied to its weld filler zone and its plate zones.

In general, the computer timesaving with the presented 2D modeling approach is significant. For example, the residual stress analysis of the single–pass weld model considered in this study takes no more than two hours for both the heat transfer and stress analysis steps. At the same time, the completion of this analysis on the corresponding 3D model takes approximately two days. The timesaving is even more significant for the three-pass weld model. The analyses times quoted above were obtained on a Windows NT workstation with a Pentium II 300 processor.

Conclusions

The possibility of replacing a computationally expensive 3D FE analysis of welding residual stresses with a less expensive 2D analysis is investigated for a single and a three-pass welded T-joint. A technique for determining welding temperature histories within the central zone of the joint with a 2D model is developed. This new technique is based on the global heat transfer balance among the entire welded joint, the section being analyzed, and the welding speed. It is demonstrated that the temperature histories within the central zone of the joint can be captured successfully by the new 2D technique. The residual stresses in the plane of the 2D model computed by this method show fairly good agreement with those computed by the 3D model. The out-of-plane stresses show about 30% difference; however, those rendered by the 2D model are conservative relative to those rendered by the 3D model. These differences can be attributed primarily to the mechanical boundary conditions in the out-of-plane direction. In a 3D model the mechanical constraint along the welding direction in the weld zone is different from the constraint in the plate zones. Such a difference cannot be accounted for by 2D FE elements.

References

Anderson, B.A.B., 1978, "Thermal Stresses in a Submerged-Arc Welded Joint Considering Phase Transformations," ASME *J. Eng. & Tech*, Vol.100, pp. 356-362.

Brown, S. B., Song, H., 1993, "Rezoning and Dynamic Substructuring Techniques in FEM Simulation of Welding Processes," *Journal of Engineering for Industry*, Vol. 115, pp. 415-423.

Friedman, E., 1975, "Thermomechanical Analysis of the Welding Process Using the Finite Element Method," ASME *J. Press. Vessel Tech.*, Vol. 97, 3, pp. 206-213.

Goldak, J., 1989, "Modeling Thermal Stresses and Distortions of Welds," in Recent Trends in Welding Science & Technology, *Proc. 2nd Intl. Conf. on Trends in Welding Research*, pp. 71-82.

Goldak J., Breiguine, V., Dai, N., Zhou, J., 1996, "Thermal Stress Analysis in Welds for Hot Cracking," *Proc. of the 1996 ASME Pressure Vessels and Piping Conference*, July 21-26, Montreal, Canada, pp. 9-16.

Gu, M., and Goldak, J.A., 1994, "Steady-State Formulation for Stress and Distortion of Welds," ASME *J. Eng. for Industry*, Vol. 116, 4, pp. 467-474.

Hibbit, Karlsson and Sorensen, 1997, *ABAQUS 5.7 Theory and User's Manuals*. Hibbit, Karlsson and Sorensen, Inc., RI.

Hibbitt, H.D., and Marcal, P.V., 1973, "A Numerical Thermo-Mechanical Model for the Welding and Subsequent Loading of a Fabricated Structure," *Computers and Structures*, Vol. 3, 5, pp. 1145-1174.

Holden, T.M., 1995, "Determination of Macrostresses (Type-I stresses) and Microstresses (Type-II stresses) by Neutron Diffraction," *Proc. Material Research Symposium*, v. 376, pp. 385-390.

Holden, T.M., Root, J.H., Holt, R.A., and Hayashi, M., 1995, "Neutron Diffraction Measurement of Stress," Physics B: Condensed Matter, v.213-214, pp. 793-796.

Holden, T.M., Root, J.H., Holt, R.A., and Roy,G., 1988, "Neutron Diffraction Measurements of the Residual Stress State of a Tubular T-joint," *Proceedings of the 7th International OMAE Symposium*, Houston, Texas, pp.127-131.

Karlsson, L., 1986, "Thermal Stresses in Welding," in *Thermal Stresses I*, ed. R. B. Hetnarski, Elsevier, pp. 300-3089.

Karlsson, R. I., Josefson, B. L., 1990, "Three-Dimesional FE Analysis of Temperatures and Stresses in a Single–Pass Butt-Welded Pipe," *J. Press. Vessel Tech.*, Vol. 112, pp. 77-84.

Lambda Research, 1996, "Finite Element Correction for Stress Relaxation in Complex Geometries," *Diffraction Notes Newsletter*, No.17.

Leggatt, R. H., 1991,"Computer Modeling of Transverse Residual Stresses in Repair Welds," *Welding Journal*, Vol. 70, pp. 299s-310s.

Leung, C. K., 1986, "Study of Finite element Modeling of Multipass Welds," Diss., University of Waterloo, Waterloo, Ontario.

Lindgren, L., and Karlsson, L., 1988, "Deformations and Stresses in Welding of Shell Structures," *Int. J. Num. Meth. Eng.*, Vol. 25, pp. 635-655.

Mahin, K.W., MacEwen, S., Winters, W.S., Mason, W., Kanouff, M., and Fuch, E.A., 1988, "Evaluation of Residual Stress Distributions in a Traveling GTA Weld Using Finite Element and Experimental Techniques," *Proc. of Modeling of Casting and Welding Processes IV*, Ed. A. F. Giamei and G.J. Abbaschian, April 17-22, pp. 339-350.

Mok, D. H. B., Pick, R. J., 1990, "Finite Element Study of Residual Stresses in T-joint Fatigue Specimen," *Journal of Mechanical Engineering Science*, Vol. 204, pp. 127-134.

Nelson, D., Fuchs, E., Makino, A., and Williams, D., 1994, "Residual-Stress Determination by Single-Axis Holographic Interferometry and Hole Drilling – Part II: Experiments," *Experimental Mechanics*, Vol. 34, pp. 79-88.

Oddy, A.S., and McDill, J.M.J., 1996, "Residual Stresses in Weaved Repair Welds," *Proc. of the 1996 ASME Pressure Vessels and Piping Conference*, July 21-26, Montreal, Canada, pp.147-152.

Pedersen, T.F., and I.L.H. Hanson, 1989, "Finite Element Calculations for Correction of Residual Stress Profiles of Coated and Uncoated Materials Measured by X-ray Diffraction," NDT International, Vol. 22, No. 6, pp.347-352.

Prevéy, P. S., 1986, "X-Ray Diffraction Residual Stress Techniques," *Metals Handbook*, Ninth Edition, Vol. 10, ed. K. Mills, Metals Park, OH: American Society for Metals, pp. 380-392.

Rybicki, E. F., Schmueser, D. W., Stonesifer, R. W., Groom, J. J., Mischler, H. W., 1978, "A Finite-Element Model for Residual Stresses and Deflections in Girth-Butt Welded Pipes," *Journal of Pressure Vessel Technology*, Vol. 100, pp. 256-62.

Rybicki, E. F., Stonesifer, R. B., 1979, "Computation of Residual Stresses Due to Multipass Welds in Piping Systems", *J. of Press. Vessel Tech.*, Vol. 101, pp. 149-154.

Sarkani, S., Michaelov, G., and Lutes, L. D., "Nondestructive Evaluation of Residual Stresses in Welded Joints with Application to Stochastic Fatigue Damage Accumulation" (accepted 2003 for publication by ASCE Special Publication).

Tekriwal, P., 1989, "Three-Dimensional Transient Thermo-Elasto-Plastic Modeling of Gas Metal Arc Welding Using Finite Element Method," Diss., University of Illinois at Urbana-Champaign.

Tekriwal P., and Mazumder, J., 1988, "Finite Element Analysis of Three-Dimensional Transient Heat Transfer in GMA Welding," *Welding Journal*, Vol. 67, p.150s-156s.

Tekriwal, P., Mazumder, J., 1991, "Transient and Residual Thermal Strain-Stress Analysis of GMAW," *Journal of Engineering Materials and Technology*, Vol. 113, pp. 337-343.

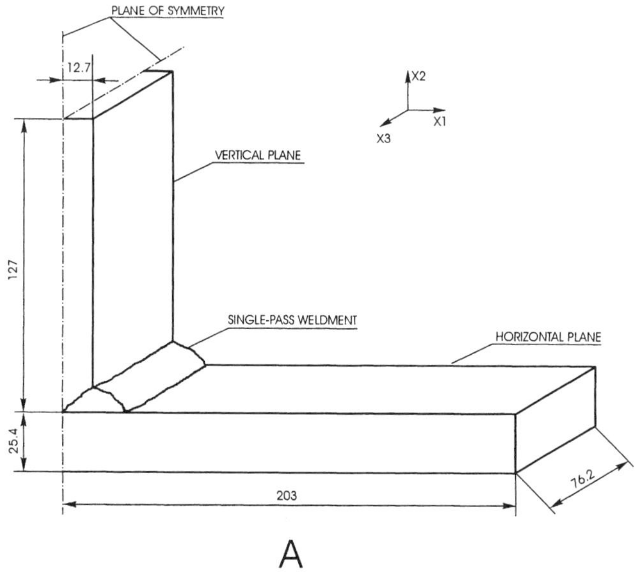

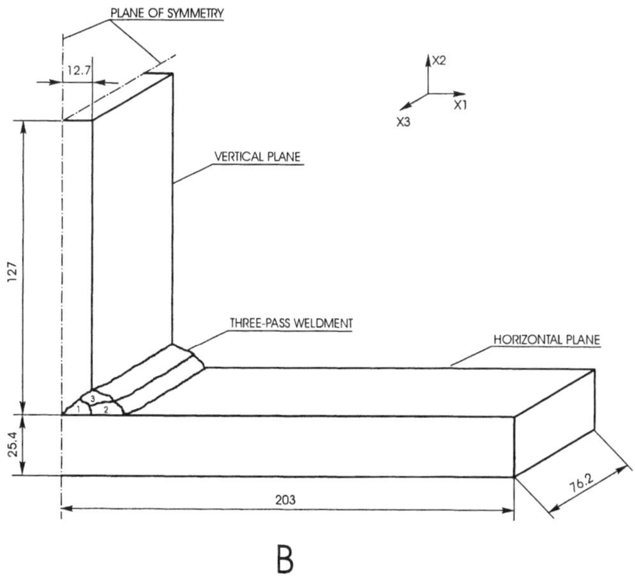

Figure 1. Geometry of the single-pass (a) and the three-pass (b) welded T-joint models.

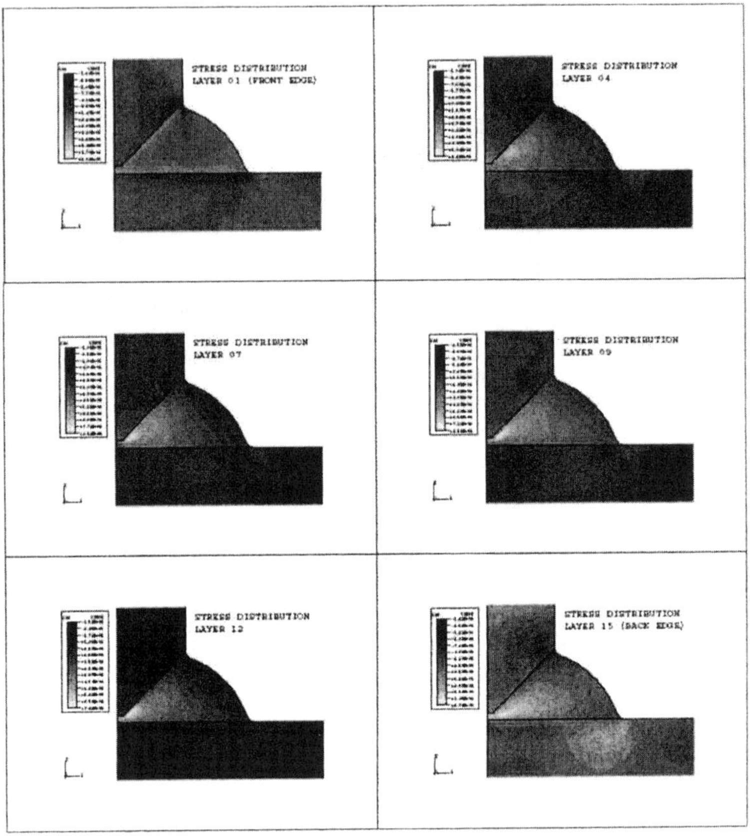

Figure 2. Variation of out-of-plane stresses in the 3D model of the single-pass weld.

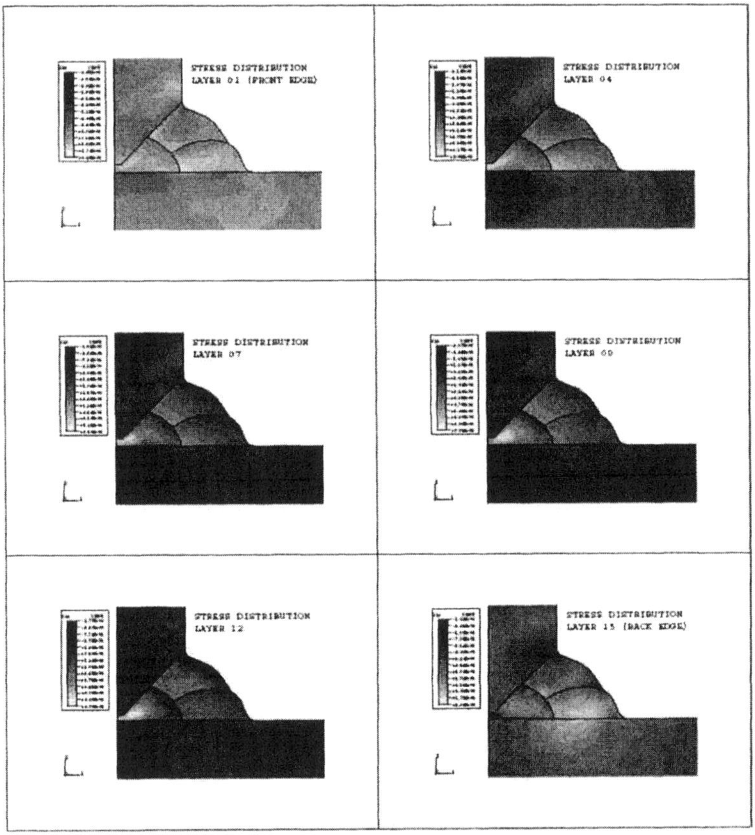

Figure 3. Variation of out-of-plane stresses in the 3D model of the three-pass weld.

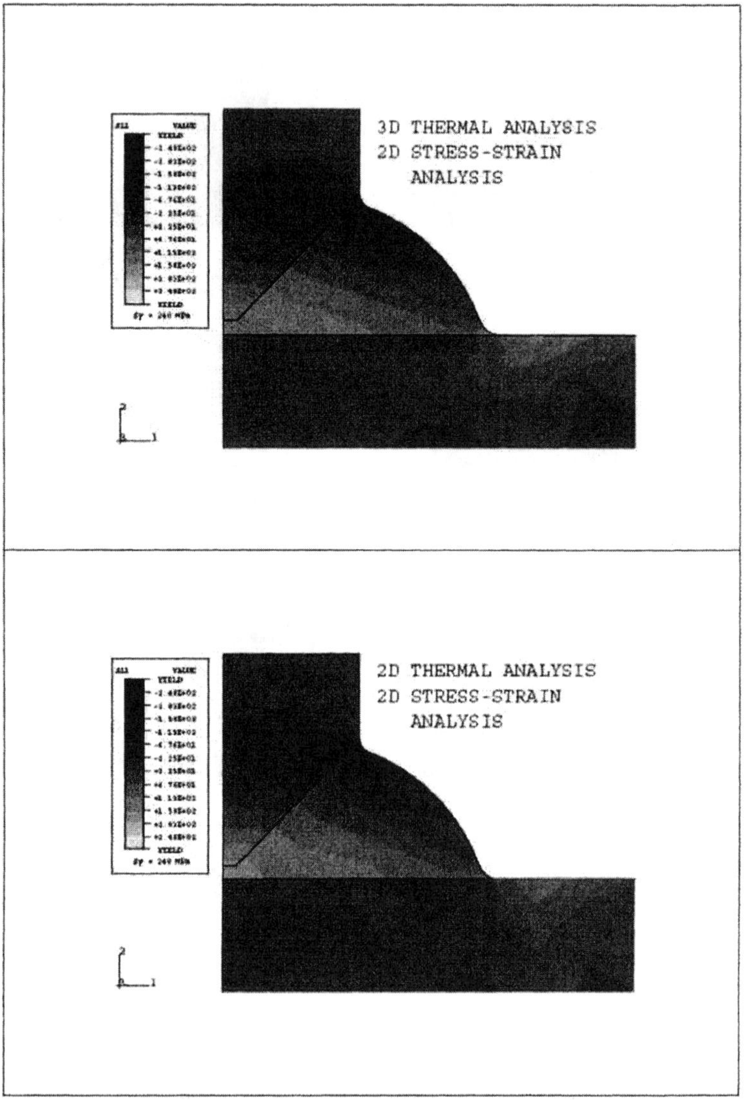

Figure 4. Distribution of residual stresses in the single-pass welded T-joint model: S11 component.

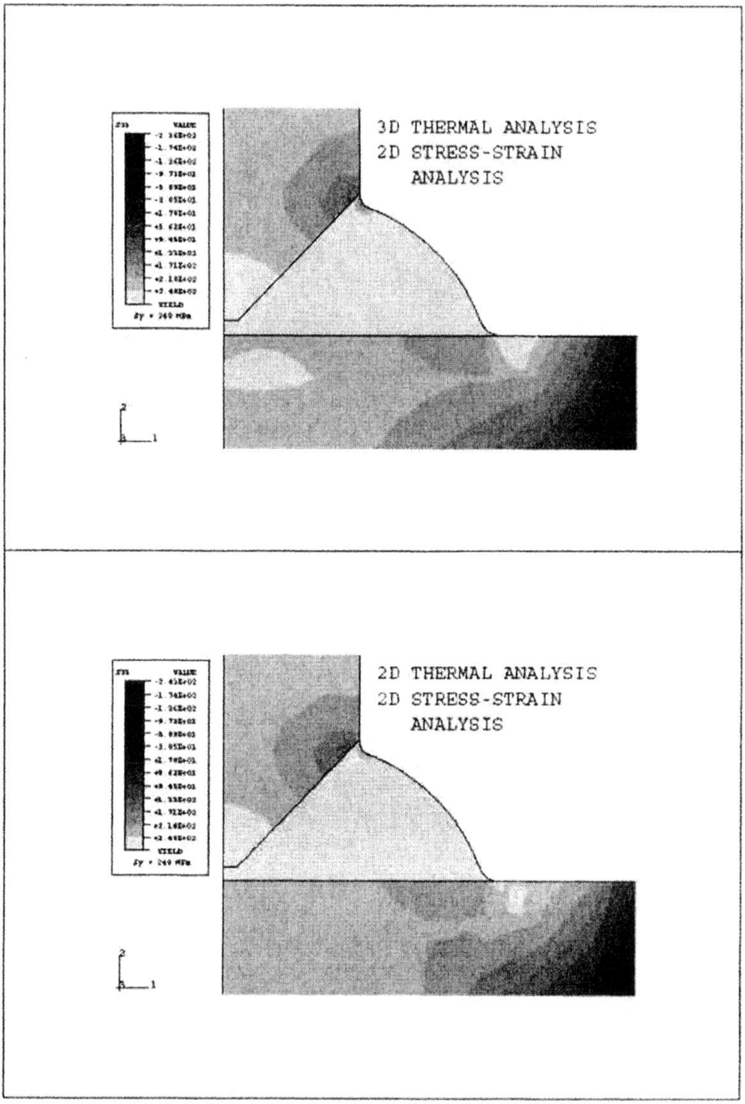

Figure 5. Distribution of residual stresses in the single-pass welded T-joint model: S33 component.

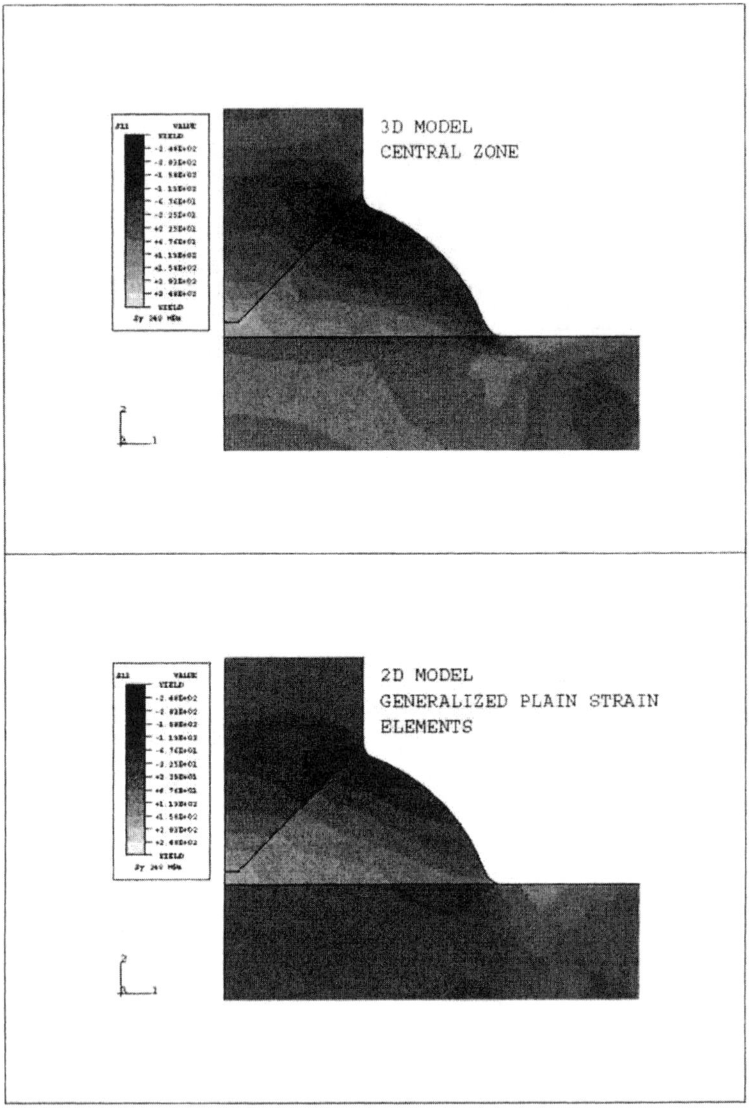

Figure 6. Comparison between S11 residual stresses obtained with the 2D and the 3D single-pass weld FE models.

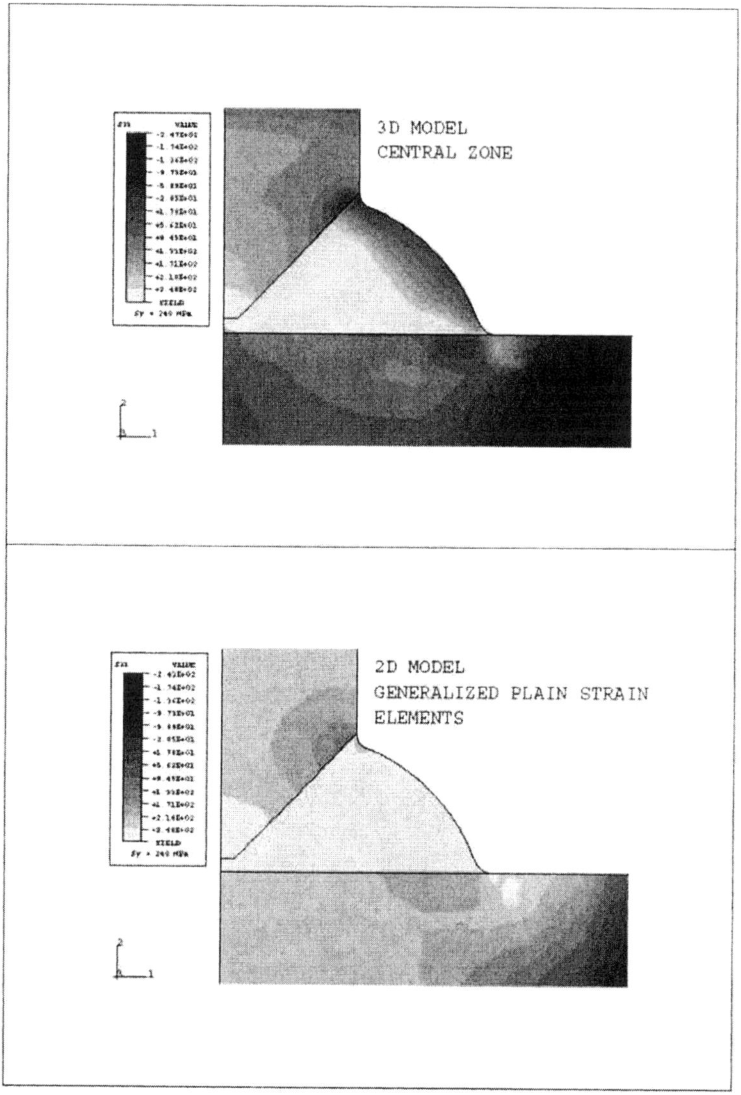

Figure 7. Comparison between S33 residual stresses obtained with the 2D and the 3D single-pass weld FE models.

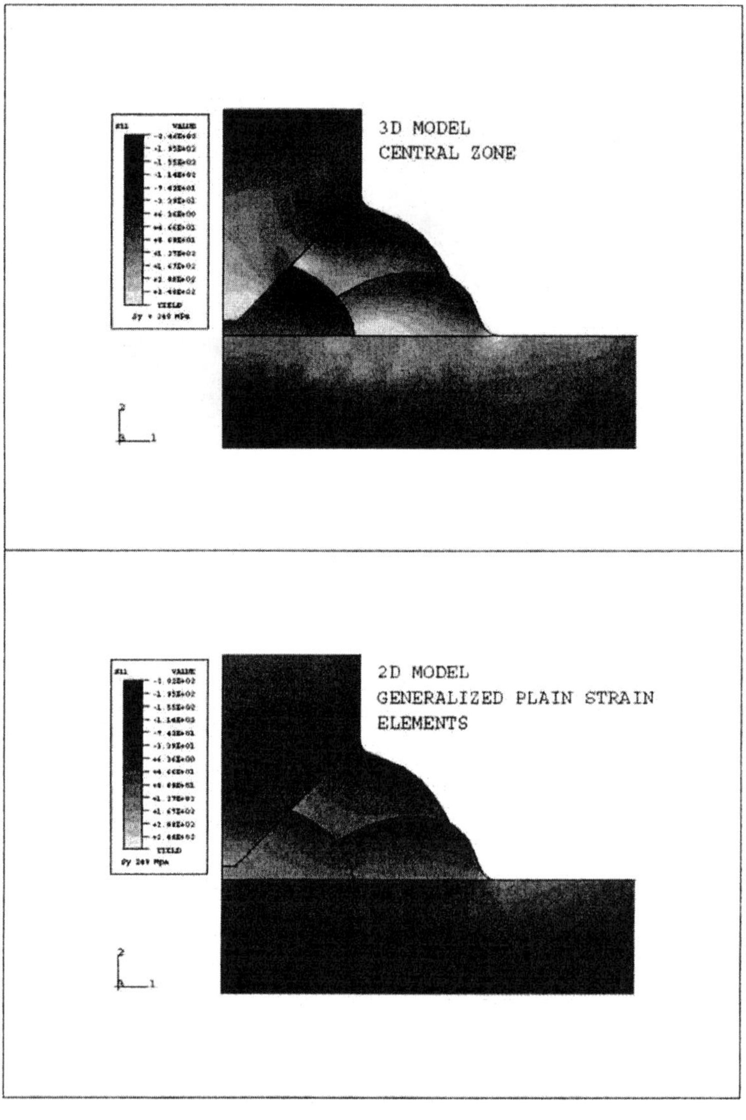

Figure 8. Comparison between S11 residual stresses obtained with the 2D and the 3D three-pass weld FE models.

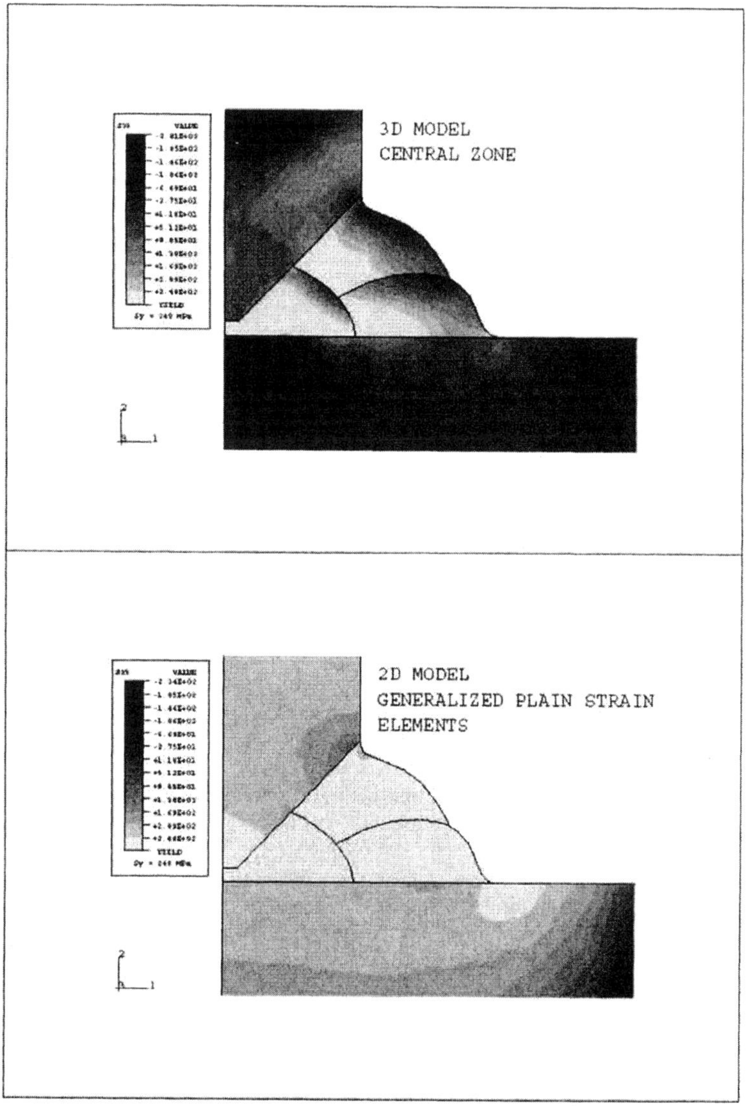

Figure 9. Comparison between S33 residual stresses obtained with the 2D and the 3D three-pass weld FE models.

A Review of Non-destructive Test Methods for Bridges

Junwon Son[1] and Jamshid Mohammadi[1], MASCE

Abstract

This paper presents an overview of various non-destructive testing (NDT) methods applied to highway bridges. The review focuses on NDT methods that are currently being used in conjunction with the condition assessment of bridge components when subject to cracking, fracture and other unseen damage cases. The review covers the availability of the NDT technology in bridge condition assessment in general.

Introduction

Structural members such as girders and slabs in highway bridges are subject to a host of loading environments that contribute to cracking and fracture. The prevalence of cracks in a steel girder of a bridge is generally due to the stress fluctuation from the passage of trucks. In certain regions, cold temperature may also trigger fracture in steel girders with or without the aggravation of the stress fluctuation effect. In reinforced or prestressed concrete girders and in concrete deck slabs, crack development constitutes a much more complicated phenomenon. Although stress fluctuation is still regarded as a major factor, many other factors may contribute to the formation of cracks and prevalence of fracture. Among these factors are the variation in temperature, vibrations generated by the traffic, operation of construction equipment at the time of repair and changes in the properties of concrete. The specific application of non-destructive testing methods utilizing data acquisition and condition monitoring for steel girders in bridges is provided elsewhere in this publication. This paper focuses on an overview of NDT methods applied to highway bridge condition assessment in general.

---

1 Graduate Student, and Professor, respectively, Civil and Architectural Engineering Department, Illinois Institute of Technology, Chicago, IL 60616

Great advances have been made in developing NDT technologies such as impact-echo, radar; induce polarization, ultrasonic, infrared, etc. to bridges. Most such methods are primarily applied to the detection of localized damage cases. However, a global condition assessment technique is desirable that can objectively quantify the overall health of a bridge and indicate any component that may suffer from damage. Local NDT methods may then be employed to the specific type of damage in the component. Generally, there is little appreciation of the defects, deterioration, loading and response mechanisms, which affect serviceability and lifetime cost. There is evidence that the design for a desirable performance is not successful at the serviceability limit state, which is characterized with cracking, deflection, and vibration. A proper application that is based on the concept of structural identification holds the key to a successful condition assessment. This paper mainly presents many different kinds of field-exploration tools for accurate and comprehensive structural identification and detection of damage conditions, many of which are accompanied by random crack patterns, fatigue and fracture of bridge components.

Nondestructive Testing Methods

Non-destructive testing (NDT) methods applied to bridge condition assessment vary widely in terms of complexity, application and cost. Simple NDT methods may only involve visual inspections; whereas, more elaborate techniques may require installation of sensors on the components, coupon testing, and use of advanced techniques such as magnetic particle testing and ultrasonic devices. In general, the NDT process starts with a visual inspection. This is an important initial step in reducing cost and in focusing on the type of methods that must be employed next. Structural monitoring using sensors is often used when information on the stress history at a critical location is sought. This information along with additional analyses is used to estimate the fatigue reliability of bridge components (Moses, et al, 1987).

Byers, et al (1997) indicate that the fatigue reliability assessment begins with an assessment of its current condition, which is a function of the service history of the structure. It is further discussed that the effectiveness of the inspection process depends on the reliability of the selected method. This reliability is described in the form of probability of detection curves or defect detection probability values. The results of a visual inspection and NDT methods are also complemented by additional structural analyses to determine the extent of fatigue crack growth in the affected structural components. As reported in Byers, et al (1997), following a repair or rehabilitation, a re-inspection program will be necessary to evaluate the effectiveness of the repair, rehabilitation, or crack control method used. The result of such an inspection can be used to update fatigue reliability and life prediction parameters through an updating process using, for example, the Bayesian approach.

NDT methods currently available have all potential applications in fatigue-critical steel components and concrete components prone to fracture and crack development. The sensitivity of each method, to a great extent, depends on the equipment and their operators. As such, their ability to detect cracks is described qualitatively instead of providing specific probability-of-detection for each method (Byers, et al, 1997).

Test methods that are currently available include: (1) acoustic emission, (2) electro-magnetic, (3) portable hardness tests, (4) liquid penetrant, (5) magnetic particle, (6) radiography, (7) residual stress by hole-drill, (8) ultrasonic, (9) Dynamic Response Signature Method, and (10) visual examinations (Iffland and Birnstiel 1993). A brief review of each method is provided next.

Acoustic emission -- Acoustic emission is defined as the high frequency stress waves generated by the rapid release of strain energy that occurs within a material during crack growth, plastic deformation, phase transformation, etc. (ASM 1985a). This energy may originate from stored elastic energy as in crack propagation, or from stored chemical-free energy as in phase transformation. Using this method, piezoelectric sensors are installed around the area where NDT is desired. By detecting stress waves generated from the energy release during the cracking process, a component damage assessment may be achieved (Lai, 1977; Collacott, 1985). The main advantage of this method is that the acoustic emission inspection detects and analyzes minute acoustic emission signals generated by discontinuities in materials under applied stress. Proper analysis of these signals can provide information concerning the location and structural significance of the detected discontinuities. The main disadvantage of the method is that it is expensive and requires complex equipment.

The Electro-magnetic method -- This method employs induced electric current as the source of the magnetic field and can be applied to ferromagnetic materials only (ASM 1985b). This method is a principal means for sorting or differentiating ferromagnetic materials on the basis of chemical, physical, mechanical and processing variables, including chemical composition, metallurgical structure, hardness and residual stresses. The advantage of this method is that it serves as a rapid, convenient substitute for conventional chemical analysis and for tests such as hardness and tensile tests, provided that reference standards of acceptability have been established and are related to indications or measurements derived from the electromagnetic test instrument. Blitz, et al (1969) and Vary (1973) report on the application of electric inspection methods in detecting active corrosion and the estimation of the depth of cracks. The main disadvantage of these methods is that they are applicable in cases where good surface contact and probe spacing can be achieved.

Portable hardness tests -- These NDT methods are primarily used to test objects that are too large or unwieldy to be tested in the usual types of testing machines. Also they can be used for testing parts of fixed structures, or for testing under any conditions, which require the indenting force to be applied in a direction other than the vertical (ASTM 1992). In order to conduct the test, surface preparation may be necessary.

Liquid penetrant inspection -- This is a nondestructive method for finding discontinuities that are open to the surface of solid and essentially nonporous materials (ASM 1985c). Indications of flaws can be found regardless of the size, configuration, internal structure or chemical composition of the work piece being inspected, and regardless of flaw orientation. A liquid penetrant is applied to the surface that is being tested. The penetrant is then removed from the surface and a developer suspension (chalk) is applied. The dye combines with the developer to produce a colored line on the surface. This will be an indication of a crack. Beyond this stage of the test, the estimate of the extent of the damage and the depth of the crack is subjective and based on the inspector's judgment (Lai 1977; Collacott, 1985). The main advantage of this method is that the liquid penetrant process is relatively simple. Equipment generally is simpler and less costly than that for most other non-destructive inspection methods. However, major limitation of liquid penetrant inspection is that it can detect only imperfections that are open to the surface. Other methods must be used for detecting subsurface defects or discontinuities. Furthermore no definite estimate of the crack depth can be made.

Magnetic particle inspection -- This is a method for locating surface and subsurface discontinuities in ferromagnetic materials (ASM 1985d). Its operation depends on whether or not the material being tested is magnetized. The discontinuities that lie in a direction generally transverse to the direction of the magnetic field will cause a leakage field to be formed at and above the surface of the part. The presence of this leakage field, and therefore, the presence of the discontinuity, is detected by the use of finely divided ferromagnetic particles applied over the surface, some of which are being gathered and held by the leakage field. This magnetically held collection of particle forms an outline of the discontinuity and generally indicates its location, size, shape and extent. A limitation of this method is that nonferromagnetic materials cannot be inspected. Such materials include aluminum alloys, magnesium alloys, copper and copper alloys, lead, titanium and titanium alloys, and austenitic stainless steels.

Radiography -- This is a non-destructive inspection method that is based on differential absorption of penetrating radiation (ASM 1985e). Because of differences in enmity and variations in thickness of the part, or differences in absorption characteristics caused by variations in composition, different portions of a test piece absorb different amounts of penetrating radiation. Unabsorbed radiation passing through the part can be recorded on film or photosensitive paper, viewed on a fluorescent screen, or monitored by various types of radiation detectors. The method is often used for inspection of weldments and detection of weld flaws. This is achieved by determining porosity, slag, and the lack of fusion penetration. The specimen is radiated with X- or gamma rays. A film placed on the opposite side of the weld will take the image. The amount of radiation that reaches the film depends on the amount absorbed by the weld (Lai, 1977). This method is slow, expensive and requires heavy equipment.

Residual stress by hole-drill method -- This method measures residual stresses near the surface of a material. The method involves attaching strain gages to the surface, drilling a hole in the vicinity of the gages, and measuring the relieved strains. The measured strains are then related to relieved principal stresses through a series of equations (ASTM 1995). This method is applicable only to easily accessible surfaces.

Ultrasonic inspection -- In this method, beams of high frequency acoustic energy are introduced into the material under evaluation in order to detect surface and subsurface flaws and to measure the thickness of the materials or the distance to a flaw (ASM 1985f). An ultrasonic beam will travel through a material until it strikes an interface or discontinuity such as a flaw. Interfaces and flaws interrupt the beam and reflect a portion of the incident's acoustic energy. The amount of energy reflected is a function of: (1) the nature and orientation of the interface or flaw; and (2) the acoustic impedance of such a reflector. Energy reflected from various interfaces or flaws might be used to define the presence and location of flaws, the thickness of the materials, or the depth of a flaw beneath a surface. (Lai 1977; Collacott, 1985). The sensitivity of this method is reduced by rough surfaces.

Dynamic Response Signature Method -- This method is based on detecting a damage scenario by comparing the dynamic response of the damaged structure with that of the undamaged one. Structural response (dynamic signature) is obtained by externally applying a dynamic load to the structure. The dynamic signature is represented in the form of a frequency response. When a structure suffers damage, changes in the signature in the form of an intensity change or shift in dominant frequencies will appear. These changes can be used to detect damage. The method has been applied to off-shore structures and in aerospace applications mainly in detecting in rotor blade misalignment in rotary-wing aircraft. The application to bridges has been reported by Rehm, et al (1987); Davis (1987), Mazurek and DeWolf (1990), Rubin (1980), and Salane amd Baldwin (1990). In general, the application of this method to civil structures has been very limited thus far. Although laboratory demonstrations have shown promising results, field applications need to be investigated further. One major disadvantage of this method is that it can become very costly. Furthermore, it often cannot pinpoint to a specific damage case.

Visual examination -- Visual inspections of the corroded metal surface is usually beneficial, and this is done under ordinary light, with or without the use of a low-power magnifying glass, to determine the extent of corrosion and the apparent location of pits (ASTM 1994). Furthermore, visual inspection can be used prior to prescribing a specific NDT method.

Nondestructive Testing Methods Used in Concrete

In addition to the methods described above, there are several other methods that are used exclusively for reinforced concrete structural members. Although these methods may not directly be used in concrete fracture reliability, they may be used as part of an overall concrete member condition assessment and their damage potentials due to cracking, corrosion of rebars and existence of voids that may further reduce system reliability against fracture and cracking.

Chain Drag/Sounding -- This is a survey technique to find areas of shallow delaminated concrete due to rebar corrosion.

Cover Meter -- In this method, magnetic fields are used to locate embedded steel reinforcement, measure depth of concrete cover and estimates reinforcing bar size. In older structures for which original drawings may not be available, this method will be effective as an initial step in identifying concrete member section properties. These properties are useful in further analysis of the section for both the strength and serviceability conditions.

Cross-hole Sonic Logging -- This is an ultrasonic test that can be used for the investigation of the integrity of concrete. Using this NDT method, such problems as soil intrusions, sand lenses, soft bottoms, cracks, breaks and even concrete quality can be identified.

Half-Cell Potential Test -- This test is used for detecting any corrosion in steel reinforcing bars. To use these methods, the concrete has to be moist.

Impact Echo -- This is a "one sided access test" to determine the thickness and condition of structural concrete voids, honeycomb, cracks, and delaminations. The test can also be used in prediction of the strength of early age concrete. It can be used in a variety of applications including in slabs, beams, columns, walls, pavements, runways, tunnels, and dams.

Induced Polarization -- In this method, an electrical current is induced into the concrete to determine the instantaneous corrosion rate of the reinforcement located below the test point.

Induction Field -- This test involves inducing an electro-magnetic field in the soil-foundation system in order to determine the lengths and continuously of reinforced concrete piles by measuring the strength of the induced field in PVC cased boring adjacent to the foundation element.

Infrared Thermaography -- This application of this method is in delamination/crack detection of concrete surfaces such as in decks, facades, etc.

Radar -- In this method, the reflections of electro-magnetic waves are measured to reveal conditions of concrete, masonry, and pavements. Using this method, metal embedded in reinforcing bars, voids beneath pavements and slabs, and regions of high moisture can be detected. One can also measure the thickness of structural members.

Rebound/Pendulum Hammer -- This method is used as a rapid survey of surface hardness of concrete and masonry. The outcome of the test can be correlated with strength. The method can be used as a quick initial step in concrete condition assessment before other more elaborate NDT methods are prescribed and implemented.

Resistivity Measurements -- This NDT method Measures electrical resistance of reinforced concrete, soil and rock to assess the potential for rebar corrosion.

Slab Impulse Response -- This is a one-sided test for slabs and pavement for determining if there exists voids or weak subgrade under footings and pavements. Furthermore, using this test one can measure freeze thaw damage depth; concrete thickness; basic material properties behind concrete walls; and locations large voids.

Sonic Echo/Impulse Response -- This is a surface test used to determine the integrity and length of a deep foundation. It is useful for forensic projects when the embedded CSL tubes are not present. Furthermore, the test can be used in verifying the results from other NDT, such as the cross-hole sonic logging, when investigating the quality of concrete in structural elements.

Spectral Analysis of Surface Waves -- In this method, the propagation of surface waves is measured between receivers to determine the variation of velocity (which depends on the material stiffness) versus wavelength (which is related to depth). Using this method, one can determine material layer thickness and stiffness without drilling for condition and dynamic property evaluations of concrete, and other cement-based materials.

Tomographic Imaging -- Tomography involves the measurement and analysis of sonic/ultrasonic wave energy over many test paths to produce an image of internal material conditions. Applications include defects in drilled shafts, and concrete/masonry structures, integrity checking of concrete dams, and sinkhole/cavity imaging.

Ultrasonic Pulse Velocity -- This NDT method is a two-sided ultrasonic test used to determine the integrity of structural concrete. Using this test, one can determine cracks, voids, intrusion, and honeycomb condition in concrete of up to 10 inches (25 mm) in thickness.

Basic Principles of Several Common Nondestructive Testing Methods

In this section, the underlying principles of several NDT methods, described in the preceding section are described.

Half-Cell Potential Methods -- When the passive coating on the steel is destroyed, due to carbonation or to the presence of chloride ions above the critical concentration, conditions are favorable for corrosion.

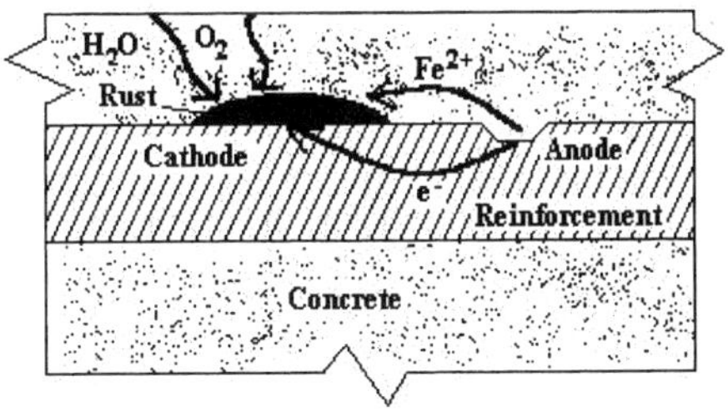

Fig. 1. Localized corrosion of steel bar embedded in concrete; Iron is dissolved at anode and rust forms at cathode

Heterogeneities in the surface of the steel, such as differences in grain structure and composition, and local differences in the electrolyte, due to the heterogeneous nature of concrete, cause a region of the bar to act as an anode and another region to act as a cathode, as shown in Fig. 1. At the anode, iron atoms lose electrons and move into the surrounding concrete as ferrous ions (Fe2+). The electrons flow through the bar to cathodic sites, where they combine with water and oxygen in the concrete to form hydroxyl ions (OH-) (Carino, 1996). The apparatus includes a copper-copper sulfate half-cell, connecting wires, and a high-impedance voltmeter, as shown in Fig. 2.

The positive terminal of the voltmeter is attached to the reinforcement and the negative terminal is attached to the half-cell. If the bar were corroding, the excess electrons in the bar would tend to flow from the bar to the half-cell. A more negative voltage reading is interpreted to mean that the embedded bar has more excess electrons, and there is, therefore, a higher likelihood that the bar is corroding.

The disadvantage of this method is that it cannot be applied to concrete with epoxy-coated reinforcement or to concrete with coated surfaces.

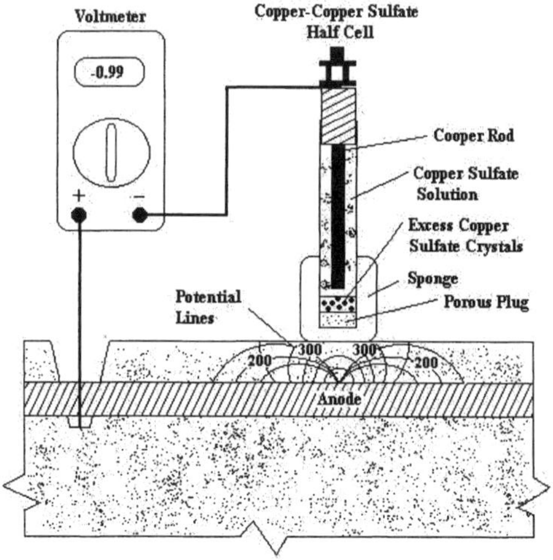

Fig. 2. Apparatus for Half-cell potential method

Polarization Resistance -- The technique basically involves measuring the change in the open-circuit potential of the short-circuited electrolytic cell when an external current is applied to the cell. There is a linear relationship between the change in voltage ΔE and the change in applied current per unit area of electrode Δi. The ratio ΔE/Δi is called the polarization resistance Rp. Because the current is expressed per unit area of the an electrode that is polarized, the units of Rp is ohms times area. The corrosion rate is inversely related to the polarization resistance (ASTM G 59).

$$icorr = B/Rp \qquad (1)$$

in which icorr = corrosion current density in A/cm2; B = a constant in V; and Rp = polarization resistance in Ω•cm3

Table 1. Relationship between corrosion current density measured and corrosion risk. (Feliú, et al. 1996; Andrade and Alonso 1996).

Corrosion current density (μA/cm2)	Corrosion risk
< 0.1	Negligible
– 0.5	Low
0.5 – 1	Moderate

> 1	High

Basic apparatus for measuring the polarization resistance of reinforcing bars in concrete is the system. The main steps for using the 3LP device to measure polarization resistance are:

(1) Measure the open-circuit potential Eo of the reinforcement relative to the reference electrode, i.e., measure the half-cell potential.
(2) Measure the circuit from the counter electrode to the working electrode that is necessary to produce a small change in the potential of the working electrode.
(3) Repeat the previous step for other small changes in potential beyond the corrosion-potential.

Impact Echo System -- The Impact Echo method involves impacting the concrete surface and recording reflections of compression wave energy with a nearby receiver for thickness measurement and detection of flaws like void, cracking and delimitation. For flat slabs and pavements without defects, a test produces a thickness echo or resonance that is used to calculate the thickness if the compression wave velocity is known. This system consists of a hand-held unit containing an electronically controlled impactor and a displacement transducer.

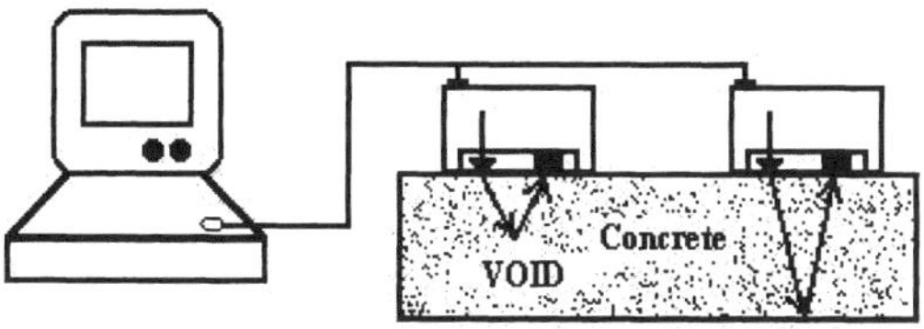

Fig. 3 Schematics of the impact echo method

Spectral Analysis of Surface Waves System -- The SASW (Spectral Analysis of Surface Waves) method uses the dispersive characteristics of surface waves to determine the variation of the shear wave velocity of layered systems with depth. The SASW testing is applied from the surface, which makes the method nondestructive. Once the shear wave velocity profiles are determined, shear and Young's moduli of the materials can be calculated through the use of simple mathematical equations. The shear wave velocity profiles are determined from the experimental dispersion curves (surface wave velocity versus wavelength) obtained from SASW measurements through a process called forward modeling or through an inversion process (Olson co., 1999).

The application of this method is primarily in: determination of pavement system profiles including the surface layer, base and sub-grade materials; determination of soil velocity profiles needed for dynamic loading analysis;

determination of abutment depths of bridges; and condition assessment of structural concrete conditions.

The SASW method requires an accessible surface for receiver attachments. The extent of the accessible surface limits the investigation depth. In SASW tests, two receivers are placed on the surface, and a hammer is used to generate the wave energy. Short receiver (typically accelerometers) spacings are used to sample the shallow layers while long receiver (typically velocity transducers) spacings are used in sampling the deep materials. Two profiles, a forward profile and a reverse profile, are typically obtained in SASW measurements where the accessible surface is struck by a hammer on two opposite sides of the receivers. A signal analyzer is used to collect and transform the receiver outputs to the frequency domain. Two functions in the frequency domain are of great importance in SASW tests: the cross power spectrum between the two receivers (used in the preparation of the experimental dispersion curve) and the coherence function (used to ensure that high signal-to-noise ratio data is being collected. The phase shift of the cross power spectrum between the two receivers is used to determine the experimental dispersion curve as follows (Olson co., 1999).

$$t(f) = FXY(f) / (360° \cdot f) \tag{2}$$

$$VR(f) = D / t(f) \tag{3}$$

$$lR(f) = VR(f) / f \tag{4}$$

where t (f) = time delay between receivers as a function of frequency, f; FXY (f) = phase shift of the cross power spectrum in degrees; VR(f) = surface wave velocity; D = distance between receivers; and, lR(f) = wavelength. The disadvantage of this method is that the SASW measurements require one surface to be accessible for testing. The depth that can be tested by SASW measurements is sometimes controlled by the accessible surface extent. A thin layer of slow velocity material lying between two thick high velocity layers cannot be identified, particularly if this layer is deep.

New Technologies in NDT

The NDT technology is a dynamic field. Many new techniques are currently stilled under development. A few such emerging technologies are described in this section.

Fiber-Optic Bragg Grating Sensor -- Current instrumentation techniques for monitoring the performance of constructed structural elements are mainly limited to the applications of electric resistance principles or magnetic principles using strain gauges, linear differential transformers, and acoustic-based instrument. Main problems associated with these systems stem from their sensitivity to ambient electrical noise and potential of degradation with age. Recent developments in the field of fiber-optic sensor technologies offer advantages that can essentially eliminate deficiencies such as the ones mentioned, and permit the long-term reliable and remote quantitative monitoring of changes in deformation of existing constructed facilities (Chen; Maher and Nawy). The theoretical relationship between the fractional changes in wavelength with the induced strain is

$$\varepsilon = \kappa(\Delta\lambda / \lambda) \tag{5}$$

in which κ = strain coefficient for the grating; the parameter λ is the light wavelength; and ε = the strain applied to the optical fiber. And the average crack widths can be calculated from the expression,

$$w'mean = \varepsilon f\,(l/n) \tag{6}$$

in which l = effective gauge length; n= number of crack widths; and εf = strain obtained from the FOBG sensor.

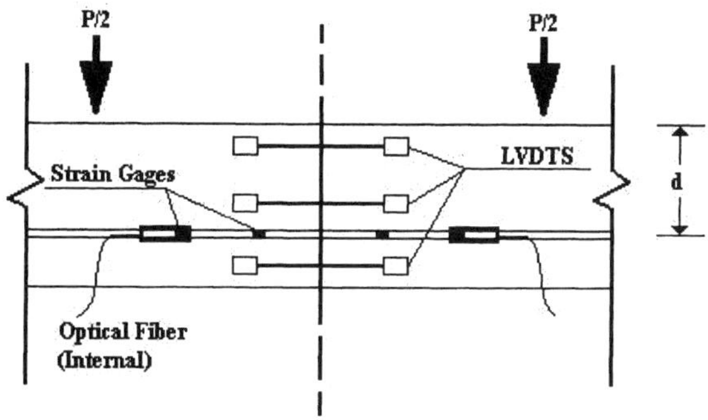

Fig. 4. Instrumentation with internally mounted FOBG sensors

The advantage of this method is that Fiber-optic Bragg grating (FOBG) sensors measure strains through a spectral shift a narrow wavelength band, which provides good isolation from noise sources such as intensity fluctuations caused by a light source or bending loss in the lead fiber. The narrowness of this band gives good resolution and stress. This sensor also can be embedded into the concrete of beam by mounting the sensor tubing (electric foil strain gauges) on the tension reinforcement, as shown in Fig. 4. Unlike LVDTs, the fractional shift in wavelength of a FOBG sensor is directly proportional to the strain induced by the applied loads rather than the relative displacement between the two measuring points.

Aerial Bridge Inspection -- Current practice involves placement of an inspector near structural components so visual analysis can be performed. The job can be dangerous, difficult and may be conducted under adverse conditions. When inclement weather, poor illumination or other impediments prevail, the inspector's ability to examine the structure may be compromised. A robotic aerial bridge inspection platform is under development that will make bridge inspections safer, faster, more efficient, and less expensive. A tethered robot platform that will fly to the inspection locations from a ground-based site is under development to automate, quicken, and improve the process.

The first phase of this project has produced a flying platform on which inspection instrumentation can be mounted. The second phase will develop a sophisticated pilot control system. The platform is capable of vertical takeoff and landing, and it will be equipped with a high-resolution video camera. The aerial platform is designed with unique, electrically driven ducted-fan technology. For added safety, there are no exposed moving parts. With this system, a camera can be positioned within 60 cm of the critical bridge component. The cable also carries control signals and video images through fiber-optic lines. The video camera can be panned and tilted to "see" from any angle. The IE-GBS, the heart of the system, consists of a computer and associated equipment with the capability for real-time display of images received from the inspection camera. The system acquires, stores, and processes images with the added capability of highlighting specific features so cracks and structural deformities may be readily identified. Structural elements that are too dark or too low in contrast to be seen with natural vision can be enhanced with sophisticated system components [Intensity Dependent Summation (IDS) hardware].

The image can be video recorded with a time stamp for later review and documentation. The advantage of this method is in

(1) enhanced images that help in locating structure cracks;
(2) improved safety for inspection crews; and
(3) image storage for review after inspection.

Advanced Ground-Penetrating, Imaging Radar -- Ground Penetrating Radar (GPR) is a technology that is becoming of importance for inspection of high-value structures, such as bridges. In a typical inspection application, inspectors use GPR to locate structural components, like reinforcing bars embedded in concrete, to avoid weakening the structure while collecting core samples for detailed inspection.

Advanced GPR, integrated with imaging technologies for use as a nondestructive evaluation (NDE) tool, can provide the capability to quickly locate and characterize construction flaws and wear- or age-induced damage in these structures without the need for destructive techniques like coring.

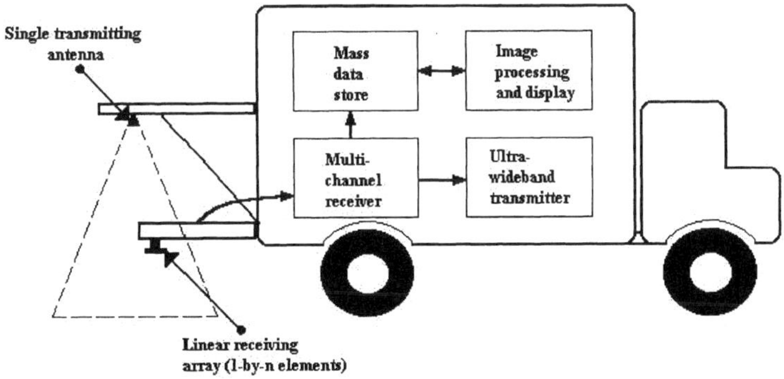

Fig. 5. GPIR bridge deck inspection concept

The bridge deck and its wearing surface are the most vulnerable parts of a bridge to damage from routine service, and they are particularly well suited for inspection using a vehicle-mounted inspection system. In an advanced bridge deck inspection system, like the one shown in Fig. 5, mobile Ground Penetrating Imaging Radar (GPIR) gathers data for high-resolution image reconstruction of embedded defects and features. High-quality images, processed in centrally located computing centers; allow visualization of internal structure, permitting evaluation of deck conditions from data that are usually obtainable only by destructive means (Warhus; Hernandez; Nelson; Johansson and Lee, 1993).

Summary and Conclusions

An overview of various non-destructive testing methods applied to bridge condition assessment of structures is provided. Various NDT methods applied to condition assessment of concrete elements are summarized and described. Most such methods are not directly used in the reliability assessment of concrete elements against fracture and crack development, however they are often necessary for the determination of the condition of the concrete and extent to which damage had progressed before one can proceed with fracture reliability evaluation. The following are the main conclusions:

Common to all NDT methods is a visual inspection of the structural elements. The visual inspection can provide a preliminary information on the type of NDT that will best suit the structure.

A variety of NDT methods can be used directly in, or in conjunction with, fatigue and fracture analysis and condition assessment of structures.

References

Andrade, C., and Alonso, C. (1996). “Corrosion rate monitoring in the laboratory and on-site.” Constr. And Build. Mat., 10(5), 315-328.

AHMCT Engineering, University of California at Davis web site. Http://www-ahmct.engr.ucdavis.edu, 1999

Aktan, A. E, et al (1999). "Condition assessment for bridge management," J. of Performance of Constructed Facilities, ASCE, 13 (3), 96-106.

ASM (1985a). Acoustic-Emission Inspection, Metals Handbook, American Society for Metals, Metals Park, OH.

ASM (1985b). Electromagnetic Sorting and Testing, Metals Handbook, American Society for Metals, Metals Park, OH.

ASM (1985c). Liquid-Penetrant Inspection, Metals Handbook, American Society for Metals, Metals Park, OH.

ASM (1985d). Magnetic-Particle Inspection, Metals Handbook, American Society for Metals, Metals Park, OH.

ASM (1985e). Radiography, Metals Handbook, American Society for Metals, Metals Park, OH.

ASM (1985f). Ultrasonic Inspection, Metals Handbook, American Society for Metals, Metals Park, OH.

ASTM (1992). Standard Test Method for Indentation Hardness of Metallic Materials by Portable Hardness Testers, ASTM, 1916 Race Street, Philadelphia, PA.

ASTM (1994). Standard Guide for Examination and Evaluation of Pitting Corrosion, ASTM, 1916 Race Street, Philadelphia, PA.

ASTM (1995). Standard Test Method for Determining Residual Stresses by the Hole-Drilling Strain-Gage Method, ASTM, 1916 Race Street, Philadelphia, PA.

Bernold, L. E., Bai, Y., Lorenc, S. J., and Carver, C. (1996). Future Technologies for Bridge Maintenance, Report Published by Civil Engineering Department, North Carolina State University, Raleigh, NC.

Blitz, J.W.G., and Rogers, D.G. (1969). Electric, magnetic, and visual methods of testing materials. Butterworth's Londom, England.

Byers, W.G., Marley, M.J., Mohammadi, J., Nielsen, R.J., and Sarkani, S. (1977). "Fatigue reliability reassessment procedures: state-of-the-art paper," J. of Struct. Engrg., 123 (3), 271-276.

Carino, Nicholas J. (1996). "Nondestructive techniques to investigate corrosion status in concrete structures," J. of Infrastructure Systems, 2(3), 108-117

Chen, Benxian, Maher, M. H., and Nawy, Edward G. "Fiber-optic bragg grating sensor for nondestructive evaluation of composite beam." J. of Struct. Engrg., 120 (12), 3456-3470

Collacott, R.A. (1985). Structural integrity monitoring, Chapman and Hall, New York, NY.

Davis, A.G. (1987). "Application of dynamic test methods to the evaluation of bridge structures." Bridge evaluation repair and rehabilitation, A.S. Nowak and E. Absi, eds., Uni. Of Michigan, Ann Arbor, Mich.

Feliú, S., González, J A., and Andrade, C. (1996). "Electrochemical methods for on-site determinations of corrosion rates rebars." Techniques to assess the corrosion activity of steel reinfored concrete structures, ASTM STP 1276, N. Berke, E. Escalante, C. Nmai, and D. Whiting, eds., ASTM, West Conshohocken, Pa., 107-118.

Iffland, J.S.B. and Birnstiel, C. (1993). Guide to Non-Destructive Test Methods, Bridge Management 2, Inspection, Maintenance, Assessment and Repair, Thomas Telford Services Ltd, London.

Lai, D.M. (1977). "Non-destructive inspection of steel (phase 10)." Rep., California Dept. of Transp., Sacramento, Calif.

Mazurek, D.F., and Dewolf, J.T. (1990). "Experimental study on bridge monitoring techniques." J. Struct. Engrg., ASCE 116(9), 2532-2549.

Moses, F., Schilling, C.G., and Raju, K.S. (1987). "Fatigue evaluation procedures for steel bridges," Nat. Cooperative Hwy. Res. Program, Rept. No. 299, Transp. Res. Board, Nat. Res. Council, Washington, DC.

Olson Instrument. Co. web site. Http://www.olsoninstruments.com, 1999

Rehm, G., Luz, E., and Bidmon, W. (1987). "Non-destructive test for early damage detection." Bridge evaluation repair and rehabilitation, A.S. Nowak and E. Absi, eds., Uni. Of Michigan, Ann Arbor, Mich.

Rubin, S. (1980). "Ambient vibration survey of offshore platforms," J. Engrg. Mechanics Division, ASCE, 106 (6), 425-442.

Salane, H.J., and Baldwin, J.W. (1990). "Identification of modal properties of bridges." J. of Struct. Engrg., ASCE, 116(7), 2008-2021.

Vary, A. (1973). Non-destructive evaluation technique guide. National Aeronautics and Space Admin., Lewis Res. Ctr., NASA Sp 3079, Cleveland, Ohio.

Warhus, J.P., Hernandez, J.M., Nelson, S.D. Johansson, E.M.and Lee, H. (1993). Ground penetrating, imaging radar for bridge inspection, Engineering Research, Development, and Technology, Lawrence Livermore National Laboratory, Livermore, California, Rept. # UCRL-53868-92.

Fatigue Reliability Updating Using Nondestructive Inspections

Zhengwei (Jack) Zhao [1] Member and Achintya Haldar [2] Fellow ASCE

Abstract

A fatigue reliability updating method incorporating information from nondestructive inspections is presented. Using a linear elastic fracture mechanics approach and the first-order reliability method, the reliability index is first evaluated for fatigue damage. The reliability index is then updated using the information from the nondestructive inspections using the Bayesian approach. The method is capable of considering all major sources of uncertainties in the fatigue damage evaluation model as well as in the nondestructive inspections. The updated reliability index is then used to make maintenance decisions to mitigate fatigue damage. Information on the initial reliability index, target reliability index and permissible maintainability index is used to implement this risk-based maintenance strategy. The method addresses the overall problem of risk evaluation, maintenance, and rehabilitation of fatigue sensitive structures by combining high technology-based inspection techniques with a sophisticated reliability-based analytical procedure. The successful implementation of the procedure presented here is expected to have a far-reaching impact on the inspection, rehabilitation, and maintenance of infrastructure. It provides a vital link between reliability-based design and risk-based maintenance.

Introduction

Several methods of nondestructive inspection (NDI) were identified in the previous chapters. Nondestructive inspections are quite appropriate for fatigue and fracture damage mitigation. Since periodic inspections are required for most essential structures, including bridges, it will be very desirable if the outcome from an

---

[1] Senior Research Engineer, Structural Integrity and Reliability Department, MS 129-73, United Technologies Research Center, 411 Silver Lane, East Hartford, CT 06108

[2] Professor, Department of Civil Engineering and Engineering Mechanics, University of Arizona, Tucson, AZ 85721

inspection can be integrated with the maintenance-related decision making process. Ideally, information not only from the most recent inspection but also from all the previous inspections should be incorporated for maintenance decisions. At present, the outcome of only the most recent inspection is used for this purpose, ignoring the information from all the previous inspections even if they are available, since procedures may not be available to address the problem deterministically. Moreover, the sources of uncertainty in the fatigue and fracture-related problems are numerous, and deterministic strategies may not be appropriate.

It is essential that first a reliability-based fatigue damage evaluation be developed considering all major sources of uncertainty. The outcome from such an evaluation can be expressed in terms of the probability of failure or the reliability index. A considerable amount of uncertainty is expected in the nondestructive inspections. The reliability index then can be updated using the information from all the previous inspection outcomes. Regardless of whether a crack is detected during an inspection or not, each inspection provides additional information and the underlying fatigue reliability needs to be updated after each inspection. The updated information on the reliability index considering all the previous inspection outcomes can be used as a decision making tool for what to do next, in terms of whether to do nothing, reschedule the next inspection to an earlier date, or repair or replace the damaged structural elements immediately. Essentially, the updated reliability index can be used to evaluate the remaining life of fatigue sensitive structures.

Realizing the potential of such a risk-based fatigue damage evaluation method, the authors proposed an approach. It is a linear elastic fracture mechanics (LEFM) -based reliability model. It incorporates the uncertainty from all the random variables involved, including initial crack size, crack initiation period, crack aspect ratio, geometry of crack tip, material properties, loading, and the uncertainty introduced during the inspection itself (Zhao and Haldar, 1995, 1997). The authors demonstrated that the method is parallel to the commonly used S-N curves-based AASHTO approach in many ways; however, it is superior to the AASHTO approach since it is capable of incorporating crack size information if available. Since the method evaluates the reliability index, it is used here to update it after each inspection, as well as considering the inspection outcomes from all the previous inspections. The updating of reliability after nondestructive inspections is emphasized here. Although the proposed method can be used for any nondestructive inspection method, the results obtained from ultrasonic inspection are used in the following discussion.

Before presenting the updating method, issues related to the limitations of NDIs, the limit state function used for the reliability evaluation, target reliability, and permissible maintainability are discussed next.

Detectability of Cracks

The current available NDI methods are not capable of consistently producing the same results under the same conditions. The results are affected by all aspects of inspections, i.e., objectives, method, acceptance criteria, and interval. For a given NDI technique, the capability of the inspection can be determined by two criteria: the ability to detect cracks with a certain level of confidence and the accuracy of crack size measurement. For a specific NDI technique, the chance of detecting a given crack size depends on many factors, such as the location, environment, and size of the crack, the resolution capability of the equipment, and the professional experience of inspector. As a result, the probability of detection (POD) has been used in the profession to define the capability of a particular NDI in a given environment. Using experimental data, several statistical models have been proposed to define crack detectability for NDIs (Zhao, 1995). Among them, the lognormal model (Harris, 1977) received more attention from the profession. The general consensus is that the effectiveness of in-service inspection is much lower than that of controlled laboratory conditions. Also, to quantify the possible human error in interpreting the inspection signal, a normal distribution model has been suggested for the measured crack size (Zhao, 1995). The uncertainty in the detected crack size can be similarly incorporated. Zhao et al. (1993) suggested that a normal distribution would be appropriate for the measurement accuracy. Silk et al (1987) indicated that errors in size measurement are not necessarily normally distributed, since a crack is more likely to be oversized and a large scatter is expected in the crack size measurement. The discussion clearly indicates that mathematical models are now available to explicitly address the uncertainty in crack detection.

During the operation of an existing structure, two additional pieces of information related to the structural configuration and operation need to be considered. They are sampling information and relation information (Ditlevsen and Madsen, 1996). Sampling information upgrades the statistical knowledge about the random variables, in terms of its distribution type and statistical parameters to describe the distribution. The relation information describes the observed functional relationship between random variables and other deterministic parameters

Critical Crack Size

The critical crack size α_c is a very important parameter in developing the limit state function required for the first-order reliability method (Haldar and Mahadevan, 2000) in the LEFM-based formulation proposed by the authors. The selection of the critical crack size depends on the criterion being used. For the fracture mechanics criterion, the critical crack size can be estimated from the critical fracture toughness, K_{IC}, if the linear elastic fracture mechanics (LEFM) approach is used, or from the J-based fracture toughness, J_{IC}, if the elastic-plastic fracture mechanics approach is used. For the serviceability criterion, the critical crack size can be defined as a specific value of design crack size from which the crack growth may be unstable. Theoretically, the critical crack size should be a random variable. However, since the crack growth rate is very high near the critical crack size, the effect of the variability of critical crack size on the fatigue failure is rather small compared to other random variables involved in the entire fatigue damage process (Yao, et al., 1986). Thus, the critical crack size can be considered to be a deterministic parameter.

Target Reliability and Permissible Maintainability

To successfully mitigate the fatigue damage evaluation process, several important criteria, e.g., the target reliability index and the maintainability index, need to be used. They are discussed briefly next.

Target Reliability

The target reliability index, β_T, is defined as the reliability index below which the structure will fail to safely perform its intended function. The selection of the target reliability index depends on the consequence of failure, public sensitivity to failure, economic constraints to achieve the desired level of safety, and the experience with past successful design and construction practices. In current design codes, the target reliability index is selected corresponding to an average risk level as long as there is no obvious need to either raise or lower the existing risk levels based on performance experience.

The selection of target reliability level is a very difficult task. There is no consensus on how to select this important parameter. It depends on many factors including the perceived risk by the users as well as by the provider. Even in the same design code, the level of safety in terms of the reliability index could be different. For steel structures designed according to the Load and Resistance Factors Design (LRFD) specifications, the reliability index for an element, e.g., a beam or column, is different than that of a weld. For the bridge structures designed according to the AASHTO Specifications, the target reliability index of redundant bridges is in the range of 2.5 to 2.7, while the target reliability index of non-redundant bridges is about 3.5. For offshore structures, the target reliability index was estimated to be 2.5 according to the API RP2A fatigue design guidelines for tubular welded joints. Moreover, for a specific design guideline, the target reliability index changes with time, as more information becomes available.

The selection of the target reliability level also depends on the type of redundancy available in the system. For a non-redundant structure, the target reliability index may be higher compared to a redundant structure. Caution should be used here, since the benefit of the redundancy in a structure decreases with the increase of COV of the load variables for an optimally designed structure (Zhao, 1995). For fatigue damage evaluation, a relatively low target reliability index may be more justifiable for a stable crack growth. Compared to the catastrophic consequence of strength failure or fracture, the stable damage accumulation in a cracked member only causes a shorter fatigue life. If periodic in-service inspections are performed, the structural reliability is expected to improve. Thus, unstable crack propagation leading to fracture can be avoided with a certain level of confidence. Considering all the available information on the subject, the authors used a target reliability index of 3.0 in all their evaluations.

Permissible Maintainability Index

Most infrastructure systems, such as highway and railway bridges, offshore structures, ships, and aircraft are required to have a mandatory in-service inspection plan, on either a periodic or a non-periodic basis. The inspection plans used by most of the industries are based on a fixed inspection schedule, determined mostly by past experience with the operation and maintenance of similar structures. In recent years, more rational inspection plans have been developed using the concept of probability and reliability. In the new inspection strategies, the impact of the inspection planning on the underlying structural integrity has been considered and the inspection interval is selected to ensure that an acceptable level of safety is maintained at all times during its operation.

As mentioned earlier, the acceptable reliability level for a structure with a regular in-service inspection and maintenance plan can be lower than that for a structure without such a plan. This is not expected to compromise the underlying risk. This is mainly due to the fact that the in-service inspection and maintenance plan will provide an opportunity to detect potential excessive structural deterioration at an early stage, and catastrophic failure is less likely. To consider the beneficial effects of inspection and maintenance, a new safety index denoted by the authors as the permissible maintainability index, β_c, is introduced. The permissible maintainability index can be defined as the reliability index corresponding to the permissible safety level considering the intensity and quality of inspection and maintenance. Conceptually, the permissible maintainability index can be established considering the acceptable risk level of structural operation under a well-scheduled inspection plan. For the structure with a periodic in-service inspection and maintenance plan, the permissible maintainability index may be lower than the target reliability index, while it can be the same as the target reliability index for the structure without a regular in-service inspection and maintenance plan. The permissible maintainability index is not intended to replace the target reliability index. Whether the in-service inspection and maintenance plan will be used or not, the target reliability must be maintained for all new structural systems. The permissible maintainability index can only be used if an acceptable in-service inspection and maintenance plan is implemented. Therefore, it is intended to be used only during the operational phase of structures.

To start the discussion, the authors propose that the permissible maintainability index could be 90 percent of the target reliability index. If the target reliability index is to be 3.0, then the corresponding permissible maintainability index will be 2.7, which is close to the acceptable risk level for the fatigue damage of highway bridges and offshore structures.

Reliability Evaluation Using LEFM Approach

The implementation of the concept just discussed requires the estimation of the fatigue reliability index as accurately as possible. Using the LEFM approach and the first-order reliability method (FORM), the authors proposed such a technique. FORM (Ditlevsen and Madsen 1996) requires that a limit state function (a functional relationship among the basic design variables for a particular design requirement) be available. For the fatigue reliability evaluation, the limit state function can be defined as:

$$L = g(Z) = \alpha_c - \alpha_N = \Psi(\alpha_c, \alpha_0) - \Psi(\alpha_N, \alpha_0) = 0 \tag{1}$$

where α_0 is the initial crack size, α_c is the critical crack size, and α_N is the crack size at the N_0^{th} stress cycle. $\Psi(\ )$ is the fatigue damage accumulation function and can be defined as (Madsen, 1985):

$$\Psi(\alpha_1, \alpha_2) = \int_{\alpha_1}^{\alpha_2} \frac{d\alpha}{\left(F(\alpha, Y)\sqrt{\pi\alpha}\right)^m} = \int_{N_1}^{N_2} C\overline{S^m}\, dN = C\overline{S^m}(N_2 - N_1) \tag{2}$$

where F(α, Y) is known as the geometry function, C and m are the fatigue strength coefficient and exponent, N_1 and N_2 are the number of stress cycles corresponding to crack size α_1 and α_2, and $\overline{S^m}$ is the mean stress range. The geometry function is used to account for the possible stress concentration, and is a function of crack size, crack shape, free surface and finite width effects, the stress gradient correction factor, and the crack tip plasticity correction factor. The mean stress factor can be defined as:

$$\overline{S^m} = \int_0^\infty s^m f_S(s) ds \tag{3}$$

For the limit state function defined by Eq. (1), the probability of failure can be calculated as:

$$P_f = P[g(Z) \le 0] = \int \cdots \int_{g(Z) \le 0} f_Z(z)\, dz = \Phi(-\beta) \tag{4}$$

where $f_Z(z)$ is the joint probability density function of the random variables **Z** and the integration is performed over the failure domain. Φ( ) is the cumulative probability distribution function of the standard normal distribution and β is the reliability index based on the FORM algorithm. The reliability index required for the fatigue damage mitigation can be evaluated using the procedure considering all major sources of uncertainty. The goal of any reliability-based fatigue damage evaluation procedure is to keep the reliability index above a pre-assigned value during the service life of a structure.

Reliability Updating Using NDI

The reliability index obtained from the model discussed in the previous section needs to be updated as information becomes available from NDIs. The Bayesian updating approach is very appropriate for the fatigue damage evaluation problem since fatigue is regarded as a very slow random process. It can be used to update the reliability index and the uncertainty in the random variables in the numerical model. Zhao and Haldar (1997) proposed such a Bayesian updating technique.

A structure is expected to be inspected several times during its lifetime to insure its integrity. Since an inspection reflects the current level of integrity in the structure at the time of inspection, the limit state function of the structure as well as the distributions of all the random variables need to be updated based on the information from an inspection. As discussed earlier, the reliability of an inspection depends on the detectability of a given inspection technique and the actual crack distribution at the time of inspection. Defining the crack detectability as α_d, the possible results of an inspection can be classified into one of the following cases: (1) crack non-detection; (2) crack detection without size measurement; and (3) crack detection with measurement of crack size.

The problem can be restated as follows. A fatigue-sensitive structural member or system is expected to be inspected several times during its lifetime. Through the kth inspection and the previous (k-1) inspections (if k > 1), information on the presence of cracks was collected using an imperfect NDI technique. The inspection outcomes were defined in terms of the three detection events just discussed. In general, the inspection results are expected to be full of uncertainty. The task is to incorporate this uncertainty-filled additional information from the NDIs into the basic fatigue reliability model.

In the context of FORM (Haldar and Mahadeven , the limit state equation represented by Eq. (1), needs to be modified corresponding to the three detection events. For the event that no crack is detected during the kth inspection, the limit state equation can then be defined as:

$$I_k = C\overline{S^m}(N - N_0) - \Psi(\alpha_{d,k}, \alpha_0) \leq 0 \tag{5}$$

The event with crack detection but without size measurement can be defined as:

$$D_k = -I_k = \Psi(\alpha_{d,k}, \alpha_0) - C\overline{S^m}(N - N_0) \leq 0 \tag{6}$$

If the actual crack size at the time of inspection is equal to the measured crack size, then the event of crack size measurement can be defined as:

$$M_k = \Psi(A_k, A_{k-1}) - C\overline{S^m}(N_k - N_{k-1}) = 0 \tag{7}$$

where A_k and A_{k-1} are the measured crack sizes at the kth and $(k - 1)^{th}$ inspections and the corresponding stress cycles are N_k and N_{k-1}, respectively. If A_k is the first measured crack size in the history of inspection, then A_{k-1} will be replaced by the initial crack size α_0 and N_{k-1} will be replaced by N_0. All the parameters used in Eqs. (5-7) were defined earlier.

For the case of no crack detection in the kth inspection, given no detection in the previous (k-1) inspections, the probability of non-detection can be defined as:

$$P_{ND,k|k-1} = P\left(I_k \leq 0 \mid \bigcap_{i=1}^{k-1} I_i \leq 0 \right) = \frac{P(I_1 \leq 0 \cap \cdots \cap I_k \leq 0)}{P(I_1 \leq 0 \cap \cdots \cap I_{k-1} \leq 0)} \tag{8}$$

Similarly, the probability of crack detection after k successive inspections is

$$P_D = 1 - P_{ND} = 1 - P\left(\bigcap_{i=1}^{k} I_i \leq 0 \right) = P\left(\bigcap_{i=1}^{k} D_i \leq 0 \right) \tag{9}$$

Considering both the detection and the non-detection events, the probability of crack detection at the kth inspection, given no crack detection in the previous (k-1) inspections, can be defined as:

$$P_{D,k|k-1} = P\left(D_k \leq 0 \mid \bigcap_{i=1}^{k-1} I_i \leq 0 \right) = \frac{P\left(D_k \leq 0 \bigcap_{i=1}^{k-1} I_i \leq 0 \right)}{P\left(\bigcap_{i=1}^{k-1} I_i \leq 0 \right)} \tag{10}$$

Model Updating without Crack Detection

For any inspection, if the actual crack size at the time of inspection is smaller than the crack detectability for a given inspection technique, the crack is not expected to be detected. Based on the limit state of crack non-detection defined by Eq. (5), the probability of failure can be estimated using the Bayesian approach and can be shown to be:

$$P_{f,up} = \frac{P(L \leq 0 \cap I_1 \leq 0 \cap \ldots \cap I_k \leq 0)}{P(I_1 \leq 0 \cap \ldots \cap I_k \leq 0)} \tag{11}$$

In order to achieve an analytical solution, an equivalent limit state function can be defined as:

$$P(I \leq 0) = P(I_1 \leq 0 \cap \ldots \cap I_k \leq 0) \tag{12}$$

Thus, the updated failure probability can be further expressed as:

$$P_{f,up} = \frac{P(L \le 0 \cap I \le 0)}{P(I \le 0)} = \frac{\Phi_2(-\beta_L, -\beta_I; \rho)}{\Phi(-\beta_I)} \quad (13)$$

where β_L is the reliability index corresponding to the limit state function of the target reliability and β_I is the reliability index of the equivalent limit state function with no crack detection; ρ is the correlation coefficient of L and I, and can be expressed as:

$$\rho = \alpha_L^T \alpha_I$$

The cumulative distribution function of the bivariate normal distribution, $\Phi(-\beta_L, -\beta_I, \rho)$ can be estimated using close form formula. The corresponding updated reliability index can be determined as:

$$\beta_{up} = \Phi^{-1}[1 - P(L \le 0 \mid I \le 0)] \quad (14)$$

Model Updating with Crack Detection

As mentioned earlier, the event of crack detection can be classified into two types. If the crack is detected but the size is not measured due to the limitations of a particular NDI, the condition can be mathematically expressed as:

$$\alpha_{N,k} < \alpha_{d,k}$$

where $\alpha_{d,k}$ represents a lower limit of detectability of the NDI instrument during the kth inspection and is a random variable because of the limitations of the equipment used.

The event of non-detection ($I_k \le 0$) and detection without size measurement ($D_k \le 0$), at a stress cycle N_k, are mutually exclusive and collectively exhaustive. Thus, according to the total probability theorem, the probability of crack detection can be expressed as:

$$P(D_k \le 0) = 1 - P(I_k \le 0) = P(I_k > 0) \quad (15)$$

The updated failure probability corresponding to the event of crack detection can be expressed as:

$$P_{f,up} = P(L \le 0 \mid D_1 \le 0 \cap \ldots \cap D_k \le 0) \quad (16)$$

where L represents the initial limit state function of the fatigue reliability, as defined in Eq. 1.

The updated reliability index can be obtained following the procedure outlined by Eqs. 11-14.

Model Updating Based on Crack Size Measurement

If a crack of size A_k is detected at the kth inspection after N_k stress cycles, the event can be expressed as:

$$\alpha_{N,k} = A_k$$

Similar to the event of no crack detection represented by Eq. 16, the updated probability of failure can be estimated by the evaluation of conditional probability, i.e.,

$$P_{f,up} = P(L \leq 0 \mid M_k = 0) \tag{17}$$

But the crack size measurement event defined in Eq. 7 is an event with zero probability, and care should be given to specify how the conditional probability is defined. Theoretically, the updating problem described in Eq. 17 can be estimated as:

$$P(L \leq 0 \mid M_k = 0) = \frac{\frac{\partial}{\partial x} P(L \leq 0 \cap M_k = x)}{\frac{\partial}{\partial x} P(M_k = x)} \tag{18}$$

Usually, great effort is needed to evaluate Eq. 18 numerically, so the use of Eq. 18 may be very limited.

If crack size A_k is measured during the kth inspection, failure will occur if the crack size increases from A_k to the critical value α_c within a given time period (N-N_k). In this case, the limit state function can be redefined as:

$$L_{red} = \Psi(\alpha_c, A_k) - C\overline{S^m}(N - N_k) \tag{19}$$

According to Madsen el al. (1986), the updated failure probability is expressed as:

$$P_{f,up} = P(L_{red} \leq 0 \mid M_1 = M_2 = \ldots = M_k = 0) = \Phi(-\beta_{up}) \tag{20}$$

where the updated reliability index β_{up} is given by

$$\beta_{up} = \frac{E(L_{red} \leq 0 \mid M_1 = M_2 = \ldots = M_k = 0)}{D(L_{red} \leq 0 \mid M_1 = M_2 = \ldots = M_k = 0)} \tag{21}$$

For one observation, Eq. 21 reduces to:

$$\beta_{up} = \frac{\beta_L - \rho_{LM}\,\beta_M}{\sqrt{1 - \rho_{LM}^2}} \tag{22}$$

high technology-based inspection techniques with a sophisticated reliability-based analytical procedure. The successful implementation of the procedure presented here is expected to have a far-reaching impact on the inspection, rehabilitation, and maintenance of infrastructure. The structural reliability updating technique provides a powerful tool for integrity management of engineering system by allocating resources optimally and reducing the risk of structural failure. It is a vital link between reliability-based design and risk-based maintenance.

References

American Association of State Highway and Transportation Officials (AASHTO), *Guide Specifications for Fatigue Evaluation of Existing Steel Bridges,* Washington, D.C. (1990).

O. Ditlevsen and H.O. Madsen, *Structural Reliability Methods*, John Wiley & Sons, 1996.

Haldar, A., and Mahadevan, S., *Probability, Reliability and Statistical Methods in Engineering Design,* John Wiley & Sons, New York, NY, 2000.

D.O. Harris, *Material Evaluation*, 35(7) (1977) 57-65.

H.O. Madsen, S. Krenik and N.C. Lind, *Methods of Structural Safety* (Prentice-Hall, New Jersey, 1986).

M.G. Silk, A.M. Stonehamm and J.A.G. Temple, *The Reliability of Nondestructive Inspection* (Adam Hilger, Bristol 1987).

J.T.P. Yao, F. Kozin, Y.-K. Wen, J.-N. Yang, G.I. Schueller and O. Ditlevsen, *Journal of Structural Safety*, 3 (1986) 231-267.

Z. Zhao, A. Haldar and F.L. Breen, *Proceedings of the Sixth International Conference on Structural Safety and Reliability,* 6 (1993) 1035-1042.

Z. Zhao, *Primary and Deformation-Induced High and Low Cycle Fatigue Reliability of Infrastructure With Updating Through Non-destructive Inspection* (Ph.D. Thesis, Department of Civil Engineering, University of Arizona, Tucson, Arizona, 1995).

Z. Zhao, and A.Haldar, "Fatigue Damage Evaluation and Updating Using Nondestructive Inspections," Journal of Engineering Fracture Mechanics, Vol. 53, No. 5, pp. 775-788, 1996

Z. Zhao and A. Haldar, *Reliability-Based Structural Fatigue Damage Evaluation and Maintenance Using Non-Destructive Inspection*, Chapter 6, Uncertainty Modeling in Finite Element, Fatigue and Stability of Systems, World Scientific, 1997.

Table 1 Summary of Data of Deterministic Variables for Reference Case

Deterministic Parameters		
Dimensions of Fatigue Detail	Weld Size, T_Z	0.6 in
	Flange Thickness, T_F	1.0 in
	Width of Specimen, W	20.0 in
	Cover Plate Thickness, T_C	1.0 in
Stress Parameter, S_0		1.59 (ksi)
Critical Crack Size, α_0		2.0 in
Target Reliability Index, β		3.0

Table 2 Summary of Data of Random Variables for Reference Case

Random Variables			
Variables	Distribution	Mean	COV
Initial Crack Size, α_0	Lognormal	0.02 in	0.50
Fatigue Coefficient, C	Lognormal	8.291×10^{-11} (in ksi)	0.63
Fatigue Exponent, m	Normal	3.344	0.10
Stress Range	Rayleigh	13.79	
Stress Modeling Error, B_0	Lognormal	1.00	0.10
Average Daily Truck Traffic	Lognormal	2,500	0.10
Crack Aspect Ratio, α/c	Lognormal	0.25	0.50
Crack Detectability, α_d	Lognormal	0.10 (in)	0.50
Measurement Error, A_0	Normal	1.00	0.25

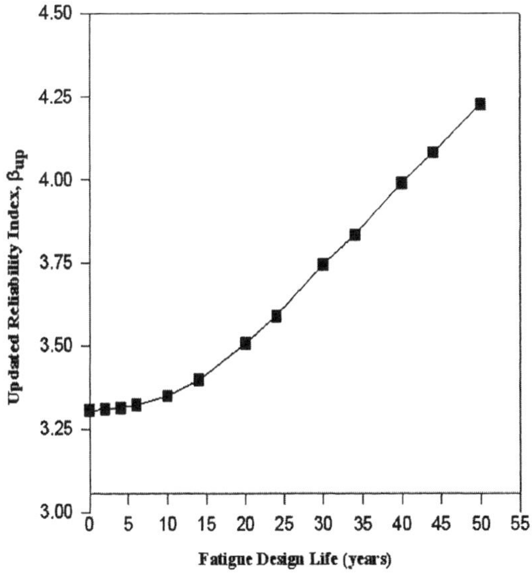

Fig. 1 Updated Reliability versus Time of Crack Non-detection Event

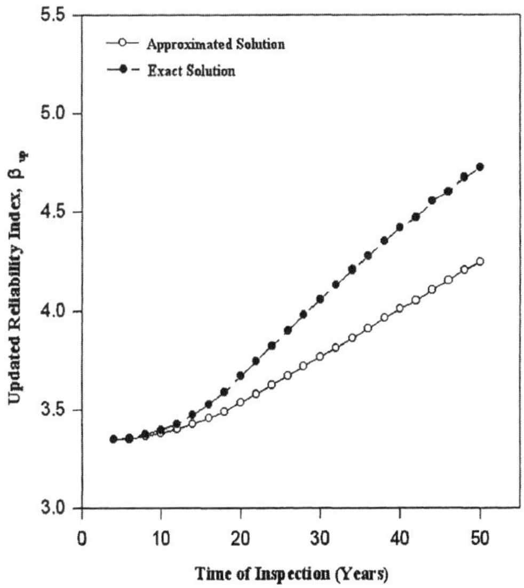

Fig. 2 Comparison of Updated Reliability Index between Approximated Method and Exact Method

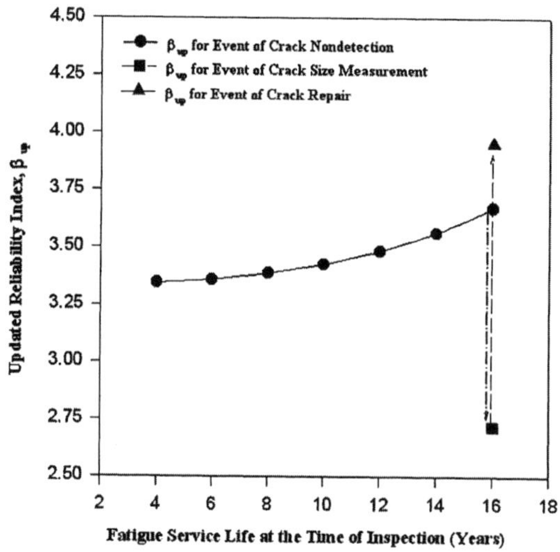

Fig. 3 Updated Reliability Indexes Corresponding to Different Inspection Events after 16 Years Life

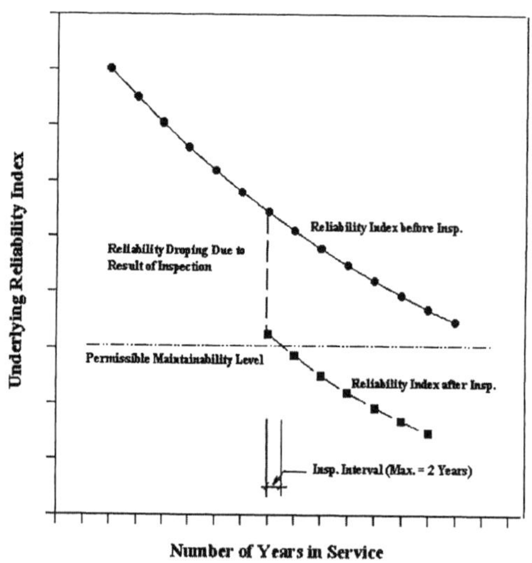

Fig. 4 Concept of Reliability Based Inspection Interval

Subject Index

Author Index